WILD CONNECTION

Jennifer L. Verdolin

WILD CONNECTION

WHAT ANIMAL COURTSHIP AND MATING TELL US ABOUT HUMAN RELATIONSHIPS

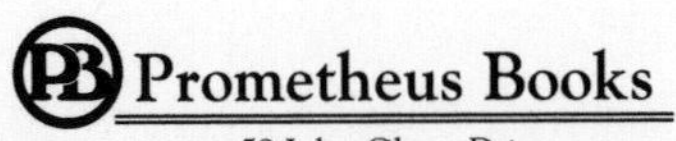

Prometheus Books
59 John Glenn Drive
Amherst, New York 14228

Published 2014 by Prometheus Books

Cover image © Media Bakery
Cover design by Nicole Sommer-Lecht
Interior illustrations created by DC Graphic

Inquiries should be addressed to
Prometheus Books
59 John Glenn Drive
Amherst, New York 14228
VOICE: 716–691–0133
FAX: 716–691–0137
WWW.PROMETHEUSBOOKS.COM

18 17 16 15 14 5 4 3 2 1

Library of Congress Cataloging-in-Publication Data

Verdolin, Jennifer L.
Wild connection : what animal courtship and mating tell us about human relationships / by Jennifer L. Verdolin.
pages cm
Includes bibliographical references and index.
ISBN 978-1-61614-946-8 (pbk.) — ISBN 978-1-61614-947-5 (ebook)
1. Sexual behavior in animals. 2. Sex. 3. Psychology, Comparative. I. Title.

QL761.V466 2014
591.56'2—dc23

2014001948

Printed in the United States of America

Dedicated to Grub (1991–2011). It was an honor and privilege to know you. You are missed.

Grub, Center for Great Apes, Wauchula, Florida, http://www.centerforgreatapes.org.

CONTENTS

AUTHOR'S NOTE

I hope that you enjoy your journey through this book. In the notes section you will be able to find all the references you'll need if you want more information about the studies used throughout the book. Additionally, you may visit my Facebook page, "What's Your Wild Connection?" at http://www.facebook.com/whatsyourwildconnection, where you can ask me questions about the animals or any of the topics discussed. You will also be able to find photos of the many different animals presented here. Lastly, I have dedicated this book to Grub, who was one of the most remarkable individuals I have had the honor of knowing. His death in 2011 left a hole in the hearts of all who were touched by his wisdom, kindness, and enthusiasm for life. I got my start working with animals at the Center for Great Apes, a sanctuary that provides a safe resting place for chimpanzees and orangutans in need of rescue and long-term care. Consequently, I have pledged 10 percent of the proceeds I receive from the sale of this book to the Center for Great Apes in memory of Grub. The sanctuary is always in need of donations, whether money or supplies. You can visit its website at http://www.centerforgreatapes.org to learn more about all the other special individuals being provided for.

1

THE BIRDS AND THE BEES

It wasn't until I was about eleven years old that I became acutely aware that there was a difference between boys and girls. Mind you, this was not a difference that I perceived, but rather, one that I was told existed. It all happened when I wanted to play little-league football. For as long as I could remember—which when you are eleven seems like an eternity—I had played neighborhood tag football. You know, the kind where you draw the plays in the sand on the edge of the road, on the grass, or even on the back of your teammate's T-shirt.

Anyway, I was told, unequivocally and without exception, that because I was a girl I was not permitted to play official little-league football. Only boys could play. This didn't make sense to me, especially since I had spent at least two years honing my football skills in the street.

I think the boys in the neighborhood got the same message, because suddenly they didn't want me to play with them anymore. The older I got, the more I noticed this pattern, and I remained perplexed. While all my girlfriends had visions of princes and castles, I had dreams of being a wide receiver in the NFL, despite the rules that clearly prevented me from participating.

The divergence didn't stop with sports. Soon after, I received a Barbie house for Christmas while my brother received the coveted erector set. Oh, how I was jealous of his erector set! Then, as I got older, the girls around me began thinking about what clothes to wear, which bag went with which outfit, and how to get the attention of that cute boy while I was trying to figure out how to climb that tree in the backyard, which mouse to play with, and how on earth I went from having two guinea pigs to six.

Into my twenties, the opposite sex remained a puzzle that I couldn't quite figure out. I knew I was attracted to men but, for the most part, I felt like (and wanted to be treated like) I was just one of the guys. As time went on, I became quite curious about this "other" sex, and I began to wonder if males were really so different from females. If they were, how did those differences influence how we interacted with each other? And why did it all seem so difficult?

Given my early adoration for and fascination with animals, my challenges in the dating world, and my questioning nature, it was only a matter of time before I found myself studying animal behavior and mating systems.

I will never forget the day in graduate school, while coteaching an undergraduate biology class, when we showed a movie to the students detailing why sex exists. I must confess that I had never pondered this particular question. I simply took for granted that we are a sexually reproducing species, and never gave it another thought.

Much to my surprise, I discovered that the evolution of sex is mysterious and the subject of much discussion among scientists. Whoa! Why hadn't this been part of my sex-education class? Like many of you, my sex-education class had diagrams of the human reproductive tract and details on why *not* to have sex. Who can forget *those* pictures? But why do we have sex? After all, not every organism reproduces sexually.

The asexual approach seems to be working well for the whiptail lizards. As many as fifty species of whiptail lizards reproduce without having sex. The appropriately named New Mexican whiptail is the official state reptile of New Mexico. I wonder if officials in New Mexico knew that the lizards never have sex before they adopted the species to represent their state.

Anyway, these little lizards can make their homes in dry riverbeds or vacant lots, eating everything from crickets to scorpions. When you reproduce by making clones of yourself, males are not required, so naturally all adult New Mexican whiptails are female. The strange part is that sometimes females do engage in "fake" sex. I know, you probably thought the strange part was that females could produce baby lizards without a male. Maybe it's just me, but I happen to think simulated sex between asexually reproducing female lizards is a tad weirder.

This fake sex pretty much follows the rules of regular lizard sex, which makes it all the more interesting. I mean, how do these females know what to do? It usually begins with one female chasing another female around, nipping at her heels. Then the female being chased assumes the donut position common to mating lizards, where curled up like a donut, she signals that she is ready. With one female pinned down by another, the one on top uses her tail to "have sex" with the other one. They reportedly thrash about, eyes closed and panting. Until now, I did not know that lizards panted. Why do some females do this? Apparently, the ones that engage in simulated copulation are stimulated into laying more eggs.[1]

Insects and other lizards, some snakes, and even a fish, the Amazon Molly, join whiptails in the pursuit of a sexless life. Okay, maybe whiptails are sexing it up just a little bit. Nevertheless, given that at least some species do not ever sexually reproduce, there must be some evolutionary advantages to not dealing with the whole boy-meets-girl thing.

One major benefit is that you don't have to share your genetic contribution with anyone else. The genes that are passed on to the next generation are, most of the time, faithful copies of the original. A second advantage is that one can skip all the wining and dining that goes into finding, attracting, and *keeping* a mate. All of this activity takes time, energy, and effort. Depending on the species, it can take precious minutes away from a very short life or years away from a relatively long one. A third plus is that one avoids all the fighting and competing over mates that is so common in the animals, including humans, that sexually reproduce.

Given what can be gained by not having sex, it starts to make sense that scientists have spent quite a bit of time trying to figure out why organisms would trade it all in. For sex to come about and then stick around, the pros have to outweigh the cons. We'll look quite a bit at the balance between costs and benefits throughout the book, since they are really what drive the development and persistence of many things, especially behavior. Clearly, since the majority of species do sexually reproduce, there are enough pros—other than the fact that it feels good—to overcome all of the real or potential costs of having sex.

How can we go about figuring out what these pros are? One clue can be found in those species that flip-flop between cloning themselves and having sex. One con is that an individual that exclusively clones itself has an Achilles' heel. It may be perfectly suited to current conditions, but as we all know, conditions change, often rapidly, and asexually reproducing species are unable to readily adapt. Why? Because strictly asexually reproducing species usually cannot produce new combinations of traits that might let some individuals survive better than others during harsh or unpredictable conditions. They lack diversity, and diversity is advantageous for long-term survival of a species.

Just as we generate entirely new dog breeds, like the Labradoodle, by combining two different breeds, sex shakes things up, shuffles the genetic deck, and creates a novel mishmash of characteristics. Variation. Diversity. Good stuff. Sexual reproduction improves the ability to fight disease, and, like I mentioned, helps individuals better deal with changing conditions.

In some sense, the advantages of sexual reproduction are like the advantages enjoyed by people who can adapt to different situations more readily than those who may be rigid and inflexible. To some extent then, those species that can reproduce either way, sexually or asexually, might have an even greater advantage. For instance, what if for some reason there are suddenly no males around when a female is ready to reproduce? Not a problem if you are a hammerhead shark, a komodo dragon, or a boa constrictor.

Hammerheads, like all sharks, are cartilaginous fishes. Until an unusual chain of events, which began in 2001, it was believed that these animals, together with mammals, were not capable of virgin births. No more. In what was like an episode of *I Didn't Know I Was Pregnant*, the staff at a Nebraska zoo was quite shocked to discover that their female hammerhead shark had become pregnant even though they didn't have any males. The initial thought was that the female, while still sexually immature, had mated with a male before being captured. Females of many species can store sperm for years. But a quick genetic test revealed that the baby female hammerheads, or pups, were exact replicas of their mom.[2] They were clones. You may love 'em or hate 'em, but the shark coolness factor just went up a notch.

For humans and all other mammals though, it looks like sex is the only way to go. This brings us to a very interesting point about humans. Except for in vitro fertilization, we need to have sex to reproduce, and since we are adding approximately 228,000 babies to the world every day, that equals a heck of a lot of sex.[3] The funny thing is that lots of people are doing it, but very few people are talking about it. And I don't mean women saying that men are thinking with their "other" head or men complaining that they can't say a word without their girlfriend or wife biting their heads off. I mean good, healthy, positive sex talk and talk about the sexes. That is where I come in. For all of you who grew up with moms who said "youknowwhat" (as one word, no hyphens) instead of penis, this birds-and-bees approach to talking about sex will be refreshing.

So let's look again at those headless boyfriends out there. Humans may use this image as a metaphor for getting their heads "bitten" off by an upset partner, but this is a real problem . . . if you are a European praying mantis. It is a myth that all species of praying mantis females, those crazy looking insects with alien-like heads, eat their mates, starting with the head. But it does happen, especially when the female is hungry. As you can imagine, males aren't too keen on being eaten.

Praying mantises are fairly voracious predators that use movement as a cue that a tasty meal is nearby. As a result, should the female look in the direction of the male as he approaches, he will become motionless, waiting for her to turn her head the other way. This is because the only reason praying mantis females chase males is to capture them and eat them. Once a male manages to get close enough to land on a female, smartly positioning his head in the direction of her tail end, he may take up to sixteen minutes before attempting to flip around and make genital contact.[4] Male praying mantis foreplay is simply trying to figure out his chances for survival.

If a male mates with a female and survives, he immediately makes a run for it by falling off of her. No sense hanging around and pressing your luck. Despite many successes, European praying mantis males fail to survive in 30 percent of these sexual liaisons. However, should the female bite off his head, this does not deter the male. He remains committed and fully capable of completing the act. He is able to do this because when she starts chewing his head off, a nerve is severed that frees the male from any inhibitions he may have had, thereby allowing him to continue mating all the way through to ejaculation. Even as the female eats her way down, perilously close to his "youknowwhat." Biology is strange indeed.

If talking about praying mantis sex doesn't break down your sex-talk barriers, at least it can make you feel heaps better about your own sex life. Not to dismiss our long list of human hang-ups and problems, but at least we can all somehow feel just a bit better that we don't have to deal with the problem of being eaten alive by our mates. It's all about perspective.

Although talking about sex is a good place to start, the impetus to write this book came from another place altogether. During graduate school I found myself in an extremely abusive relationship. Fortunately, after I graduated I was able to get out of the situation, and after a few months of freedom I began to take stock of my life, specifically the tragic dysfunctional nature of my relationships. Here I was, a woman with a PhD in evolutionary biology studying the social and mating behavior of wild animals, but for the life of me I could not figure out how dating, courtship, and "mating" were supposed to work in my own life. I became convinced that I was behaving in a maladaptive fashion when it came to love. Now, when I use the word *maladaptive*, I mean dysfunctional, not productive, or just plain not suited to the environment or circumstances.

Given that, I decided I needed to take a long, hard, honest look at myself. Talking to friends about all my hardships and confusion, it seemed that I was far from the only person who had more questions than answers when it came to dating, mating, and/or finding Mr. or Mrs. Right. Was there even a Mr. or Mrs. Right to be found? And what do you do with him or her when you find them?

When I thought about it, I realized that I knew the ins and outs of the mating behavior of the animals I studied, but I knew very little about my own species or even about *myself*. Then one evening, while questioning everything I had been doing in my "love" life, it suddenly dawned on me that finding the answers might just take a little, okay a lot, of research. It was time to get down to business and look at human problems (including my own) through a biological lens.

I decided to put all my years of studying wild animals to use in my dating life to see what would happen. In other words, maybe I could learn something about my love life from birds, squirrels, monkeys, and other animals. I figured that since you don't see female baboons or prairie dogs bemoaning how they can't get a date, wandering around confused about whether a guy really likes them, or going to great lengths to embarrass and humiliate themselves for the attention of yet another unavailable mate, maybe, just maybe, there was something I was missing. So I hatched a plan to conduct an experiment and do some "field" research. After all, as a scientist, my inquisitive mind doesn't turn off when I leave the field site or lab for the club or bar.

My effort to collect data presented me with a potential downside common to animals and people—experiencing rejection. Out in the wild, there is a lot of rejection being doled out and received in the mating game. And sometimes giving it can be just as bad as receiving it. I think the similarity ends, however, with how we humans *feel* about and *interpret* rejection. This reminds me of Koko the gorilla. As you will discover throughout this book, my first love is gorillas, and Koko the gorilla, currently forty-two, has always been one of my favorites. I remember seeing a documentary about how the researchers training Koko in sign language wanted to find a mate for her. She had a companion, Michael, but perhaps because they were raised together, their relationship had never blossomed.

Her caretakers, taking a decidedly modern approach (at least for the late 1980s/early 1990s), presented Koko with videos of suitable males. Something about Ndume must have spoken to Koko, and she picked him. Unfortunately, Ndume did not reciprocate her romantic desires. As I recall from the documentary, Ndume rebuffed her advances, even running from a frustrated Koko.

It is impossible to say for sure whether Koko suffered emotionally from this clear-cut rejection. What we do know is that she and Ndume developed a close friendship that has stood the test of time, even though they never mated. I figured that if

Koko could handle rejection, so could I. This had not always been the case.

In my early twenties I had a boyfriend, Ryan. A techie who worked for Motorola and was a huge fan of Bob Dylan, he was totally cool in a relaxed hippyish way. The fact that we had nothing in common had no influence on my dating him. Instead, I was suddenly a huge Dylan fan, loved tie-dye, and was eager to learn about techie things. This cartoon caricature of myself worked—for a while. After three months, he abruptly told me that he needed "a copilot in life," and I wasn't it. Just like that. Naturally, I vigorously objected and spent the next several weeks twisting myself even further into impossibly complex knots trying to fit his version of what a good copilot would be.

Needless to say, despite my sad and transparent efforts to be everything I thought he wanted, it didn't work. I was devastated. I had become the perfect copilot for him. I had sacrificed everything in my life to become *his* everything, and now he didn't want me? Sounds dramatic, but I really was not a Dylan fan, so, to me, I had suffered for many months listening to his daily Dylan marathons. After the deep wounds of an unrequited "love" wore off, I realized that maybe Ryan had done exactly the right thing. What I learned from that experience was that it is infinitely easier to accept the rejection and move on than to continue trying to be something I am not.

Thus, like Koko, I was prepared to put myself out there, face the possibility of rejection, and take one for the team. The human team, that is. Along the way, I examined the data on everything from just what we find attractive to our dating rituals, including cooperation, conflict, and monogamy (or lack thereof). What I found was that there were undeniable similarities between what I knew happened in animals and what happens between humans.

Suddenly, the parallels of my experience in the "field" were too compelling to ignore. This is what pushed me to write this book with myself as a guinea pig. I have come to realize that we cannot deny our link, our very wild connection, to our biological heritage.

I say this not to ignore or diminish the contribution of our complex psychology or to shatter whatever mysterious random forces are at work in our romantic pursuits. Rather, I propose that if we fail to also consider the role biology plays in driving our behaviors, we are missing a big piece of the picture. Ultimately, our biology connects us to the animal world, and I believe embracing this knowledge can only serve to enhance our efforts and experiences.

When I embarked on this experiment, I had no idea how much it would actually change my life. As I spoke to more and more people about my "findings," and my decision to write a book about my journey, I was genuinely surprised by the enthusiastic response everyone had to the idea. Besides sharing stories (some of which I have used in the book), they wanted to know if animals did this or that, or had the same problem getting their partner to cooperate or to remain faithful.

As I went along and shared tidbits of whatever section I was working on, many of my friends and acquaintances were astonished by some of the similarities we have with a suite of species. And these similarities sparked interesting conversations on even the most sensitive of topics!

What I discovered (aside from a slew of things about myself) was that not only were there peculiar and interesting similarities in human and animal behavior, but that talking about animal behavior made it much easier to talk about human behavior. Strangely, discussing cheating albatrosses, choosy peahens, or sex-crazed tortoises makes a perfect segue into talking about some very common human problems.

So, no more dancing around the subject (though dancing is involved). Throughout this book you will discover the role biological factors, often unconscious, play both in the attraction we feel toward a potential love interest and in the courtship rituals with which we are so familiar.

Parts of this book will challenge many of the traditional ideas we have about dating and relationships. Looking at how animals overcome the hurdles of communication or cooperation can add a new dimension to your romantic endeavors, whether you identify more with ducks or with albatrosses. Hopefully, by the end of this book you will be asking yourself the question I started with: What *is* my wild connection?

2

FIRST IMPRESSIONS

It's a Wednesday evening, the middle of the week, and a perfect night for a first date at the local coffee shop. You know the drill. Coffee is cheap, it doesn't take too long, and if things don't go well you haven't lost too much time or money. In walks my date. I don't remember what his face looked like, or his name, because all that remains is the image of a grown man wearing a dirty tank top, ripped Bermuda shorts (do people still wear those?), and flip flops. I don't think he had washed or combed his hair, and we will absolutely not discuss his feet. I was polite, I sat there, and I spoke to him. My mind kept wandering off. . . . *Do I have to buy cat litter? Where did I put my dry cleaning tag? Wow, that tooth is really angled funny.*

It could have been that he really had nothing interesting to say, or it could have been that no matter what he said, he would have never overcome that first impression. I made my escape within twenty minutes and was left wondering, was this guy serious? Was this the best he could do for a first date? If we go with the idea that nothing is going to be as good as the first impression, that people try to put their best foot forward and then it's all downhill from there, I could not even begin to imagine what would have been in store for me down the road.

Naturally, that got me questioning whether I was being too judgmental. Do I care too much about what someone looks like? Do I read too much into how someone presents him- or herself to the world? *Don't judge a book by its cover*; *all that glitters isn't gold*—those idioms we grow up hearing over and over again from well-intentioned parents, relatives, teachers.

Regardless of where you heard them, they were most probably accompanied by the notion that to judge the value of something or someone on appearance alone is not only wrong, but also misleading. Yet, like me, you may have noticed that despite our efforts to resist making superficial judgments, we all have that seemingly inescapable tendency to do just that—especially when it comes to people. Even with our complex intellect and all our assertions to the contrary, we assess each other, at least initially, on things we can see (and smell).

Across cultures people tend to agree on what is "attractive." Attractive people get better jobs, make more money, have more friends, and have more dating opportunities. And believe it or not, it isn't just people. It happens quite a bit among our wild counterparts. In birds, female blue tits, for instance, are better parents to the offspring they had with sexy males. Not only that, but if the male they have mated with has his coloring dulled, the equivalent of being made less attractive, the female will actively reduce her efforts to feed their offspring.[1] Why is this? Is it because we (and blue tits) are superficial and obsessed with physical appearances? Does it come from a lack of desire to value things of substance? Or perhaps there is a more elemental reason that is beyond our conscious control. So, did I unfairly judge my coffee date simply on the cleanliness of his tank top, or was there something much deeper going on that none of us are even aware of?

WHO CARES WHAT I LOOK LIKE?

How long do you think it takes you, on average, to determine the aesthetic appeal of something like a website? You may be shocked to learn that it takes only fifty milliseconds![2] And when it comes to faces, you need only a tenth of a second to make a judgment on an individual's attractiveness. But it's even more amazing than that—not only do we decide if we find someone attractive in a split second, we also come to conclusions about other characteristics, such as likeability, trustworthiness, competence, and aggression in the same amount of time.[3] Once a first impression is made, and it is *always* made, there is a long-lasting effect. In marketing and other fields this is known as the *halo effect* or *confirmation bias*.[4] Therefore, if someone makes a great first impression, you will be more likely to ignore or minimize potentially negative future issues or problems. This explains a lot doesn't it? Essentially, coming to conclusions about a person's attractiveness based on appearance alone occurs extremely rapidly, and it's largely out of our control, which suggests an underlying biological connection.

Three distinct parts of the brain light up when we meet someone for the very first time: the amygdala, the posterior cingulate cortex, and the thalamus. All three are part of the limbic system, the collection of brain structures that are involved in long-term memory, olfaction, emotions, and behavior.

Personally, I have a love-hate relationship with the first part, the amygdala, which is a little almond-shaped structure located deep in the temporal lobe. The amygdala regulates things like your fight-or-flight response, emotion, hunger, and your interpretation of socially important cues like facial expressions. It also sends out tentacle-like projections to the hypothalamus and other brain structures that process pheromones, which are critical social hormones that include the hormones involved in sexual attraction.

The second system playing an integral part in this process is the posterior cingulate cortex. This deep central region of the brain is involved in assessing value or rewards in uncertain or risky situations—like meeting someone for the first time! These two parts of the brain, the amygdala and the posterior cingulate cortex, are then linked together through the thalamus; it's about the size of a walnut and sits in the midline of both sides of the brain. It acts somewhat like a main switchboard and plays a role in everything from learning to emotion and arousal.

Given that our brain, like that of other species, works at lightning speed in making judgments about others based on appearance, the question is why? When it comes to social interactions, it is possible to imagine all sorts of benefits. For example, the faster you can decide whether someone is a friend or a foe, the less at risk you are from a potential enemy. Deciding who is competent or, if you are a bonnet macaque monkey, which males have the highest rank, can tell you who to recruit as an ally.

That's all fine and good, you say, but where does mate selection fit into this equation? There's a Buddhist saying that it takes six months to get to know someone and any judgments on character should not be made before the six months are over. Since it doesn't seem like there's any life-threatening hurry in choosing a mate within the first fifty milliseconds of meeting him or her, what's nature's rush to get us to make life-altering decisions in much less time than it takes to wink at someone?

It boils down to this: when it comes to attraction and mate choice, it is all about finding a high-quality mate, and our biological wiring is designed to help us connect with the best possible match. Appearance is the fastest and most direct initial link to this information. As you will see, physical appearance is so critical to mate assessment in animals that it is relatively immune to cheating.

PICKIN' A GOOD ONE—A MATE, THAT IS

What constitutes a good mate varies from species to species. For some species, it could come down solely to healthy genes (you don't see a male bear sticking around to help mom raise the little ones), while for other species being a strong protector or a good provider matters just as much. One feature many bird species have in common is that, oftentimes, both parents are needed to successfully raise the clutch. If you and your partner happen to be mourning doves, perhaps you take shifts, roughly equal in length, sitting on the eggs to incubate them, and later feeding your nestlings. Emperor penguins probably have the toughest job, since they absolutely must rely on their partner. Living in one of the harshest environments on the planet, the Antarctic, with frequent blizzards, temperatures averaging -20 degrees Celsius in the winter, and winds ranging from 60 to 125 miles per hour, successful mating requires Herculean cooperation and commitment by both males and females. Because their breeding grounds are far from their food source, both parents fast for long periods, with the males bearing the brunt, fasting for as long as four months.

During courtship and mating the male and female solidify their bond. The female then transfers her one and only egg to the male, after which she heads off to find food and fatten up before returning. Males will protect this egg, balancing it on their feet, almost to the death. Men may joke about women always being late, but here the price for tardiness is steep. Males do occasionally abandon eggs or chicks, but this happens only if they absolutely cannot continue without food—usually because the female is very late in returning. Once the female does come back, the male transfers the chick to her and is free to go find food. Then he returns and she goes. This cycle continues until they have raised their chick. For emperor penguins, choosing a low-quality mate is disastrous. They only have one egg and one chance per mating season.[5]

So, what had Mr. Flip Flop's first impression really told me about him? Was he subconsciously sending me evolutionary messages that he'd leave me holding the egg? Or maybe that he wasn't necessarily the sharpest tool in the biological shed? The other question that begs for an answer is this: With so much on the line, is a tenth of a second really enough to be assured that your partner is up to snuff, a good mate, provider, or protector who will contribute equally to the chores if that is what is required in your species?

WHAT BODY PARTS ARE THE MOST REVEALING?

Since we already know that appearance is a major deciding factor, what physical traits give the most information? If you're a red-legged partridge, the color of your eye ring may hold the key. I know, you probably thought I was going to say the legs. I mean why else have red legs? But no, in this case, beauty is not in the eye of the beholder so much as, well, in the eye.[6]

Carotenoid pigments, the chemical compounds that produce yellow, orange, and red coloration in many animals, including people, are not produced by the body on its own. Instead, the colors are the result of diet. For instance, if you consume enough carrots you will develop an orange hue fairly quickly. You are what you eat, literally. In the red-legged partridge, the deeper the hue of these colors around the eye of the male, the better he is at finding food. This in turn tells the female, indirectly, that her choice is a high-quality male because he can successfully find food. More importantly, redder males have fewer parasites and a stronger immune system. The end result: female partridges go gaga over deep red eye rings. And when they do get down with a really attractive male, they lay more eggs and invest more energy into the eggs they produce.

We may think that the eyes are the windows to the soul, but long before you've even thought about the soul, you've already decided whether or not those windows are attractive to you. I'm guessing you've probably never thought about just how similar you are to a female partridge. For us humans, it is not so much the color of the eye, but rather the size and presence of the limbal ring around the iris.[7] If you go to the mirror and look at your eye, you will notice that your iris, the colored part of your eye, has an outline or ring around the edge. This is the limbal ring. The darker and more prominent the ring, the higher the attractiveness score. This cuts across genders, with both males and females finding opposite and same-sex individuals with thicker limbal rings more attractive.

Unfortunately we cannot eat our way to a better limbal ring. You are born with the ring you have, and over time it gets thinner and thinner. Why would we care about something as small as a ring around the iris? As it happens, this little ring can tell you a lot about the health and age of a person. It is affected by several medical conditions, such as unhealthy stem cells in the corneosclearal limbus (the whites of the eye), corneal adema (swelling), and glaucoma. It also degrades over time so that the older you are, the thinner the ring becomes. If you are like most people, you are probably not consciously aware of it and most likely have never even heard about it before this very moment, but you can bet that your subconscious is paying close attention to it every single time you decide how attractive someone is.

But back to the birds . . . and their diets. As I already mentioned, a poor diet leads to dull or absent coloration. This can be in the eye, as is the case of the red-legged partridge, but in many other bird species it is more often the feathers and beak that hold the key to attractiveness. Most female birds have a preference for certain traits in their partner, which, in turn, influences the appearance of males.[8]

Peacocks illustrate this perfectly. If you look at a male and female side by side the male is the flashy one and the female, or peahen, is drab by comparison. Since the time of Darwin and his book *On the Origin of Species*, the peacock has been the iconic symbol of female choice. Yet the symbolism of the peacock goes much farther back in time. In Catholicism, the peacock represents immortality, renewal, and resurrection and is depicted in third-century Roman catacombs and in paintings such as Fra Angelico's fifteenth-century nativity scene. In Buddhism, the peacock is associated with wisdom, and it is represented in several Hindu deities. But alas, not all peacocks are created equal, and modern research has revealed just how finicky peahens really are.

The male peacock's elaborate, ornamental train enamors us all as he fans it about, trying to impress everyone from females to other males. Within the train there are iridescent eyespots decorating over two hundred feathers. When the male isn't prancing around showing off his tail, the long train of feathers is dragged along the ground. While having to drag your feathers along the ground may be annoying, there is a heftier price that males pay for lugging around all those feathers. It is the effect this drag has on their flying ability. The weight of their tail compromises their lift, leaving them at greater risk for attack by predators. Another cost is the energy that must go into producing such an elaborate and colorful tail. After a lot of research and head scratching by scientists about why peacocks have such a costly ornament, findings show that it's because peahens like them. Ha! That's right, just because the girls fancy them.

Specifically, females prefer a male with a lot of eyespots on his tail. Peahens mate less frequently with males that they previously found attractive but whose eyespots had been removed. More interestingly, peahens don't just up and mate with the first male with a lot of eyespots that comes along. Instead, females choose after visiting an average of three males. One study found that in ten out of eleven cases the males that were chosen had the highest number of eyespots. So are peahens running around counting eyespots? And was female number eleven just not good at math? This is where things get really amazing. Females consider males with over 150 spots to be super sexy, but peahens aren't counting. Instead, they are looking at the symmetry of the left and right side of the tail, and males with the most spots also happen to be the most symmetrical![9]

So, symmetry, or the similarity between the left and right side, matters in judging attractiveness. But is this fancy with symmetry restricted to peahens? Definitely not. Although the Mediterranean fruit fly, or medfly for short, is not a very exciting creature, except maybe for its propensity to destroy fruit crops, it also shows a preference for symmetry in mate selection. Males of many fruit fly species, including this one, have complex courtship rituals in which they "sing" to females by flicking their wings around. Males have a series of, shall we say, unique steps they must go through. First, a male must establish a territory—a place where he will do all the necessary medfly stuff needed to convince a female to mate with him. He carefully chooses the underside of a leaf. Once he's got his turf staked out, he sends out a scent, or pheromone, from a special gland to let the females know he has arrived. We will talk more about these important pheromones soon, but for now let's see what happens when a female comes to his leaf. First, he fans his wings, exposing his pheromone gland, perhaps trying to ensure that she gets a full whiff of him. So many analogies come to mind, but let's stay focused. Next, he starts flicking his wings back and forth and rocking his head from side to side. Assuming she is mesmerized by his fabulous display, he moves closer and closer until he jumps over her head, lands on her abdomen, and tries to mate with her. So much for foreplay! Of course, if she becomes disenchanted with him at any point during this process she just flies away.

With all the smells, sounds, and dancing going on, where does symmetry fit in? My first thought was that it must be the wings, but I was wrong. My second idea was that it could be the male's bright white chin. Who knew fruit flies had chins? But no, it isn't the chin either. Rather, it is the symmetry of two bristles on the top of their heads, shaped like spatulas and called supra-orbital-frontal bristles, or SFOs (why can't we just say hairs on top of the head?) that mean a lot to finicky female fruit flies. Females are so picky that over 90 percent of mating attempts are rejected.[10] Is there a purpose to these spatula-shaped hairs? Some researchers think that they are there to get the female to pay attention to his bright red eyes, while others think the hairs help him get his scent closer to her. Funny how the male goes through all these physical gymnastics and, in the end, it's the symmetry of the hairs on top of his head that makes all the difference in the world.

Peacocks and medflies, okay, but what if you are a three-spined stickleback? This little fish makes a big splash when it comes to science. From differences in appearance depending on location to their elaborate mating behavior, three-spined sticklebacks have long been of interest to evolutionary biologists. They get their name because of the three spines located on both sides of their body just in front of the rear, or dorsal, fins. When males are trying to impress females they erect their spines. Of course they do! But even more impressive is that the males defend territories and take care of the eggs and eventually the little fry, or fish babies.

This is actually pretty common in fish, where males do the bulk of the work. One reason for this is that fish eggs are externally fertilized; that is, the female deposits the eggs externally and the male then covers them with his sperm. As a result, the female can lay her eggs in a male's nest and take off before the male fertilizes them. She basically leaves him holding the bag. As I was saying, males court females by erecting their spines, but females don't seem to care how long the spines are. Instead, they really care about how perfectly similar the spines are on the left and the right side of the body. And like their very distant cousins, the medfly, male three-spined sticklebacks with more mesmerizingly symmetrical spines convince more females to lay more eggs in their nests.[11]

I could go on endlessly about the varied ways that symmetry plays a prominent role in attraction, but I'll stop after just one more example. When Japanese macaques were presented with photographs of faces that had been digitally altered so that one image was perfectly matched on the left and right side (100 percent symmetry) and the other was adjusted so that the left and right side were only half identical (50 percent symmetry), both males and females gazed longer at the more symmetrical faces, although this preference was stronger in females.[12]

What about humans? Whether the face is real or computer generated, or even the face of a macaque, people generally rate symmetrical faces as more attractive.[13] Even more interesting is that when women are the most likely to get pregnant, based on where they are in their menstrual cycle, they have a much stronger preference for facial symmetry as well as for more "masculine" voices and faces.[14]

What about men? Do they care about symmetry? Not surprisingly, they do. And like women, it is not just about symmetry, but also about how "femininely" symmetrical a woman's face looks. Females with more feminine facial features (for example, dainty chin, small nose, narrower jaw) have greater fertility and reproductive success.[15] Facial symmetry can be so intoxicating that people with more symmetrical faces have more sexual partners; and when people cheat, they tend do so with people who have faces that are highly symmetrical.[16] This is not why they cheat, mind you; we'll talk about that later. Even children as young as nine start showing a clear bias for more symmetrical faces![17] So why all the fuss about the left and right side of *everything* being similar?

Although it is still controversial, the similarity between the left and right side of your face, your wings, your bristles, or your eyespots may be tied to how healthy you are. While we're looking at a face, thinking, *Wow, she/he is really beautiful*, that beautiful, highly symmetrical face is telling us that the body is also highly symmetrical and the genes that built that body are pretty darn good. And perfect symmetry is not easy for nature to reproduce. It requires the combination of many biological and environmental factors, meaning that the more perfect one's symmetry is, the less imperfect one is, at least at a biological level.

For example, in both people and chimpanzees, greater asymmetry, or larger differences between the left and right side of the face, is related to poorer health.[18] In the case of people, this includes self-reported health on things like respiratory infections, number of times antibiotics were used, and intestinal infections, as well as actual health records. In a captive group of chimpanzees, health measures came from keepers and veterinary care records. In people, facial asymmetry may not be the only indicator of poor quality. It turns out that greater asymmetry indicates lower fertility and semen quality in men.[19] The same has also been found for a wide range of animals from antelopes to insects.[20]

The basic idea is that we would all be perfectly symmetrical but for some kind of disturbance during development: a bad environment, a faulty gene, a disease, or something else that creates an imbalance. More importantly, only some individuals, perhaps those with a stronger immune system or better genes, can withstand the forces that would make us all imperfect.

What does all this mean? It means that facial symmetry, and subsequently attractiveness, is an honest signal about a potential mate's physical health and fertility. And that is why we all pay attention to it. Does this mean that we don't choose a mate for her superior intelligence or his excellent conversational skills? No. It simply means that we are wired to see symmetry as beautiful, and this plays a predominant role in the *immediate* attraction we have toward someone.

The repercussions for being a bit crooked can go beyond just poor mating success. It can have dire consequences. Suppose you are a little wood mouse minding your own business, gathering some food, preparing your nest, or grooming yourself. Suddenly you detect a predator, maybe a tawny owl, but you are caught off guard and your only choice is to run. If your legs are not the same length on the right and left side of your body you may be out of luck. Why? Well you won't be able to run as fast or in exactly the right direction. The physics of locomotion mean that the greater the asymmetry, the less evenly body mass is distributed along its axis. Bad for the wood mouse, good for the tawny owl.[21] I suspect that if a human were trying to escape from a predator, he or she would run a bit faster, too, if both legs were the same length.

NOSES, TEETH, HAIR, FEET, OH MY!

What about other traits? There is this notion that there is a universally attractive male and female type, especially with respect to the face. Despite the fact that some have argued that people prefer physical features that are closer to the average of a population, meaning that very large or small eyes, ears, or noses are not deemed to be as attractive, some people do fancy extremes, like large noses or chins.

Two of my friends come to mind. One likes the long, strong "Roman nose," while for the other friend, any large nose will do. Perhaps they would fall madly in love with the proboscis monkey found in Asia. In this species, the male's nose is substantially larger and longer than the female's. It hangs lower than his mouth, like a phallic symbol, enticing the female. Does this penchant for big noses have something to do with perceived masculinity?

I don't know about noses, but this is apparently true for chins. There are differences among men and women in chin shape and size, and some women really like broad chins because they think it gives a more masculine impression.[22] For men, large eyes, small noses, and dainty chins are the hallmark of facial attractiveness because they tend to be associated with femininity.[23] Even other men perceive men without broad chins as weak. I think this might explain the remark, "He has a weak chin." As mentioned earlier, females with more feminine facial features actually do have greater fertility and reproductive success. So, you see, these are not simply random preferences. They are tied to our genes, the strength of our immune system, and our fertility.

I must confess that I am obsessed with teeth. There, I said it. And for me it crosses genders. Thankfully I am not alone. I have discovered along the way that many people pay tremendous attention to teeth and find a good set of chops particularly attractive.

Think about what teeth tell you about a person. How a person's teeth develop is linked to their genes and the kind of environment they grew up in. As an adult, the state of your teeth is an undeniable record of your environment, traumatic events, disease, and aging. Not to mention your bank account! Over a billion dollars are spent annually in the United States alone on cosmetic procedures to make one's teeth look better.

I went on a blind date with a man who told me beforehand that he'd had an accident that affected his teeth but that he lacked the money to fix the problem. He clearly was concerned and self-conscious. Given my fixation with teeth, it may very well have been an insurmountable obstacle had I liked him otherwise. Fortunately I did not, though I am fairly certain he was convinced it was because of his many missing teeth.

The desire to improve the appearance of teeth is universal, and if you have enough money you can fix almost anything, well, except your genes. So in the absence of cosmetic procedures to disguise them, the state of a person's real teeth acts as yet another honest signal about mate quality.[24]

Another confession. . . . I really like my hair. It is shiny and soft, and I have a lot of it. Apparently other people like it, too, as strangers have a tendency to touch it. I have been in line at the grocery store, minding my own business, when out of nowhere I feel hands running through my hair. Yes, complete strangers have touched my hair without permission. Quite frankly, it's a little creepy. I wonder if this is how pregnant women feel when people touch their stomachs without waiting for an invitation to do so.

Does my hair provide a signal about my health? Apparently it does. Healthy, shiny, strong hair is linked to good physical health.[25] After a woman has children, her hair tends to suffer, becoming darker, drier, and brittle. The same thing happens with age. Hair length can be important, too, as many men prefer long hair. Perhaps this is because hair is an inescapable log of the current state of your health and diet. The longer the hair, the longer the history.

The value of hair in terms of mate attraction is a two-way street. When surveying my friends on what they found attractive in a mate, several females positively swooned at the thought of a man with thick, wavy, shiny hair. Male hair quality and quantity is also linked to health, and studies show that male pattern baldness is linked to a higher risk of coronary heart disease.[26]

From the time of Aristotle, it was believed that bald men were more virile and "manly." What you might not know is that they may have been on to something. Though there is no evidence for linking hair loss with greater fertility, hair loss in men is associated with overall higher testosterone levels.[27] This is genetic and not linked to the amount circulating at any given moment in the bloodstream, which is precisely why women are much less prone to lose their hair. Though most women might be attracted to a man with a full head of hair, don't tell that to my friend Charlotte, who swears that balding men are the sexiest by far. Though psychologists can probably have a field day with this one, from a biological standpoint, it means that Charlotte (whether or not she consciously knows it) likes her men with healthy doses of testosterone.

What about hair color? Does it matter? It does for lions. Even though I personally think all male lions have swagger, lionesses are a bit pickier. They unapologetically prefer males with fuller, darker-colored manes.[28] As with many species, including humans, a darker mane distinguishes a male who has more testosterone coursing through his veins, which also makes him more aggressive. Even male lions pay attention to the mane color of their rivals.

Among the big cats, lions have a complex social structure. Females form lifelong stable groups, or prides, made up mostly of relatives. Males are transient, leaving the group they were born in to search for a group of females to control. When young males leave, they may do so in the company of other young males (brothers, cousins, and so forth). This gang, or coalition, of lions may be just a pair, or it could be made up of as many as ten males. The job of male lions is to look handsome and defend their pride of females at all costs, whether they are alone or in a group. When researchers place stuffed lions into a male's territory—one with a dark mane and one with a light mane—the male usually approaches the intruder with the lighter mane color. This is probably because the territory holder deems lighter-colored males less aggressive and easier to get rid of.

If you play the sound of a female lion roaring, dark-haired males lead the way, trying to be the first to get to the female. And it usually pays off because lionesses mate with dark-haired males more often. Essentially, males with darker manes, and all that accompanying testosterone, are more aggressive and more successful in fighting, and they produce more babies and more babies that survive to adulthood.

Like the lionesses on Africa's Serengeti Plain, human females are more often attracted to a male with a good head of dark hair. Men with darker hair are judged as more attractive and more intelligent, reliable, mature, and masculine. For human males considering females, it tends to be the reverse. Blonds may indeed have more fun, since men generally rate fair-haired women as more beautiful.[29] Perhaps this is why almost 40 percent of US women highlight, lighten, or dye their hair? Lucky for me, some men do prefer brunettes!

Speaking of hair, what about facial hair? Because facial hair is associated with masculine traits, I think it is safe to assume that facial hair on a woman could be perceived as unattractive, but what about on men? My friend Rachel is distrustful of men with facial hair, and she finds full beards particularly distasteful. She's not sure if it is because she thinks such men are hiding something or because she just generally dislikes body hair. After all, we differ from all other primates in that we have become relatively hairless—well, at least most of us.

Researchers have determined that men with facial hair are perceived as having a high social status and as being older and more aggressive. However, what they are not is more attractive. Of course some women do like men with beards and other extremely masculine facial traits (for example, a pronounced brow ridge), but this may be linked to the man's perceived social status, age, and ability to provide more than just physical attractiveness. This seems particularly true in countries with lower healthcare standards and economic stability.[30]

What this means is that although women do not, on average, find men with full beards more physically attractive, they do rate them as more powerful or dominant. Dominant males are perceived as having better access to resources, which may explain why, in areas of low economic stability, women favor men with beards. Males also rate males with beards as being more aggressive and having higher status. This might explain why the beard never really goes away, regardless of cultural pressures influencing humans at any given time. For example, there is archeological evidence of shaving as early as ten thousand years ago, suggesting that removal of facial hair has a long history.[31]

However, beards were a status symbol for the Egyptians, though beards again lost popularity when Christianity emerged and labeled the beard as demonic. In eighteenth-century England, being clean-shaven represented being mannered, refined, genteel, and open. The beard made a comeback in the nineteenth century, and Abraham Lincoln helped the cause with his famous facial hair. It then went out of fashion yet again when people began to focus more on personal hygiene. The beard seems to have become fashionable yet again, but don't be fooled gentlemen, women and men may respect you more with a full beard, but they won't find you more attractive.

Just as facial hair and beard preferences change with time, so do preferences regarding feminine curves. Marilyn Monroe, then Twiggy, and now, once again, feminine curviness has reappeared as sexually appealing. Though, to be fair, for some men the appeal of a narrow waist and full hips was never lost. Across cultures, from New Zealand to Brazil, a lower waist-to-hip ratio is universally deemed most attractive.[32] More importantly, this is independent of weight. Break out the cheesecake, ladies, because this means that some men find women with an hourglass shape extremely appealing.

Now you might be saying, "Ok, but how is that news?" The news part of it comes when we consider why this is so. And it isn't just because men arbitrarily like hefty bosoms and ample rumps. A woman's waist-to-hip ratio is tied to her age, hormone levels, fertility, and susceptibility to disease.[33] For instance, women with large breasts and smaller waists have higher estrogen and progesterone levels, hormones linked to fertility. But big breasts don't seem to do better when it comes to producing more milk, and there doesn't seem to be a clear preference by men one way or the other.

The real deciding factor regarding female body attractiveness is this waist-to-hip ratio, with a sweet spot of around 0.6 to 0.8. In the three to four seconds that men are actually paying attention to a woman's head and body shape, they are primarily fixated on the breasts, even though the waist-to-hip ratio is what really determines attractiveness. It seems men discern pretty rapidly how narrow a woman's waist is and then spend the remaining 3.8 seconds looking at a woman's breasts or butt, if that is available for view.

Perhaps the way the buttocks look from behind has something to do with why many men like women who wear high-heeled shoes.[34] In heels, the hips tilt back a smidge and they sway more seductively, not to mention the more feminine way a woman walks when in heels. Well, some women. An ample rump is not only attractive to human males, but also to male baboons. Though to be fair, the large swelling on the rear of female baboons is not really their rump. It is called the anogenital region. The swelling of this region coincides with a female's estrus cycle, becoming more prominent during ovulation—and the larger the better.[35] The larger the size of a female's swelling, the more likely she is to conceive. Hail to the rump!

Although every single possible physical trait that we might find attractive is beyond the scope of this book, one "fetish" deserves special attention, namely feet. Other than a movie starring Eddie Murphy, I was completely oblivious to the attraction some people (mostly men) have to feet. That is, until I embarked upon yet another blind date. We had exchanged photos, and a few days before the date we talked on the phone. It was at this point that the guy, says, "Your picture looks very nice, but I have to be honest, it doesn't matter to me how beautiful you are if you have ugly feet." Huh? "So," he continued, "would you mind sending me a picture of your feet?" What? I must say it took me a moment to respond. You expect a man to ask for pictures, but not necessarily of your feet. I readily agreed. I was fascinated. The scientist in me reasoned that it would be interesting to see how my feet fared. The woman in me was scared by the thought that I might have ugly feet. I always thought my feet were fine, kind of pretty, you know, for feet. I had given up obsessing about other parts of my body only to find out that now I might have to worry about my feet!

Though Footman and his fetish were quickly history, I continued to wonder, where does this passion for feet come from? Neuroscientists may have provided us with the best possible explanation for those who feel irresistibly drawn to feet. In mapping how the brain processes sensory perception, studies have revealed that the brain regions that process sensation from the feet and toes sit right next to the regions that process sensory information from the genitalia![36] While this may not have any bearing on fertility, and it may not reflect mate quality in any way, it's clear that good-looking feet are an important trait to some . . . which, once again, brings us right back to my coffee experience with Mr. Flip Flop.

THE CHEMICAL WORLD: DOES HE PASS THE SNIFF TEST?

As if your face, your eyes, your symmetry, your hair, your waist, your teeth, and sometimes even your feet weren't enough, there's even more going on than meets the eye, and this other consideration may just be the ultimate deciding factor in initial mate attraction—it's how you smell. When discussing mating systems with my students I always tell them, "It wasn't that you saw each other across the room; you smelled each other!"

There is one man I know who I could probably smell a mile away. And I mean that in a good way. For me, his natural scent is so thoroughly intoxicating that I can barely think when I am around him. Worse still, I am like a Basset Hound when it comes to sensing his presence in my environment, and I am convinced it is because I smell him before I see him!

The idea that one can become inebriated by the natural smell of another is not as strange as it may sound. Both males and females of many species succumb to the scent of desire. Just look at the delightful, pudgy, brown lemming male: he has a keen sense of smell when it comes to the ladies. It's a rough life for lemmings. They are pretty much on the bottom of the food chain, and they only live for about a year and a half. Not one for dilly-dallying, this little rodent packs a lot into that short life span.

Despite living in the arctic, lemmings are active all year round. With the clock ticking down, there is no time to hibernate for these guys. Females can have several litters a year, raising anywhere from four to nine babies at a time. Females have the uncanny ability to sniff out better mates, and their noses lead them right to the dominant male. Have you ever walked into a room and said, "Boy, you can just smell the testosterone in the air?" Apparently this is what female brown lemmings are discussing as well.

Male lemmings not only have a knack for smelling females that are ready to mate, but also for smelling those that haven't already mated with another male.[37] When females mate with multiple males, it is harder for males to be sure of their paternity. Male lemmings try to get around this by detecting whether a female has already mated. This, of course, implies that males leave a chemical calling card that other males can detect. From beetles to bees and lizards, females do give off a different scent if they have already mated or if they are ready to mate.

What does all this chemical calling card stuff have to do with us? Lo and behold, we are just as sensitive to the scent of the opposite sex as the humble lemming. Humans can discriminate odors in just a single whiff, which at a minimum takes approximately four hundred milliseconds. Like male beetles, bees, lizards, lemmings, and a whole suite of other species, men can discern the scent of a woman ready to become pregnant. They find the smell of sweat from women who are close to ovulation more pleasing and even sexier.[38] And not just their body odor, men also prefer the voice, the complexion, and basically everything about a woman near ovulation. The thing is, men know women are ovulating because they can smell it, but they don't know that they know!

One of my friends swears by this phenomenon. She claims that she gets a lot more attention from men right before she begins ovulation. Whether it is holding the door open for her, buying her a cup of coffee or a drink, or being asked out, like bees to honey the men flock to her, only to disappear again once she passes that magical time. Women, the same holds true for us. When we are ovulating we strongly prefer the scent of a male, but not just any male, a more symmetrical male (there is that darn symmetry again!).

Beyond the simple fact of whether one prefers certain scents, there is increasing evidence that how an individual smells, the person's pheromone signature, if you will, may be linked to that person's genetic health—specifically, his or her immune or disease-fighting genes. These are known as major-histocompatibility-complex (MHC) genes. By distinguishing at a cellular level between self and other, they are involved in identifying and fighting off invading pathogens.

Mothers, fathers, and close relatives like grandparents and aunts and uncles have been shown to be able to identify the odor of a related infant compared with an unrelated one.[39] In the case of fathers and other relatives, they can do this even if they have had no prior exposure to the baby! When we look to animals, we find similar results. Individuals seem to be able to tell the difference between relatives and nonrelatives based on smell alone. And it is largely thought that this is due to the scent one gives off based on the particular set of MHC genes you have.

While this is fascinating—and potentially a topic for another book—what does this have to do with finding and choosing a mate? Studies with lab mice reveal that, all other things being equal, individuals will choose a mouse mate that is most dissimilar in the MHC genes.[40] This phenomenon extends far beyond the lab. One of the cutest species I have had the pleasure of studying is the grey mouse lemur. This nocturnal primate, native to Madagascar, is small enough to fit in the palm of my hand, reminding me of a miniature Topo Gigio, an Italian television puppet character popular when I was growing up.

Looking at grey mouse lemurs in the wild reveals that their mate choices are also MHC-dependent.[41] The benefits of this are twofold. First, they avoid mating with relatives, and second, by combining different genes from two parents, offspring have the maximum diversity in their disease fighting genes. This second benefit may help offspring survive better when fighting off infections and disease.

I mentioned that mothers and other close relatives can distinguish the smell of a related versus nonrelated infant, but does this extend to detecting the best genetic match based on MHC composition? Yes indeed. Just like paper wasps, the house mouse, seabirds, primates, and countless other animals, human females have a stronger sexual interest in the odor of males who differ from them on the MHC-gene level. Even more interesting is that in already-paired couples, women were less sexually responsive to and had fewer orgasms with partners who had similar MHC compositions.[42] Perhaps this is why some men are constantly obsessed with whether or not a woman has an orgasm? As if that weren't bad enough, closely MHC-matched couples also engaged in a higher number of extrapair copulations.[43] Translation: *more cheating*.

Similarity of MHC composition may also explain why some couples have difficulty getting pregnant, and it may even explain the frequency of spontaneous abortions.[44] With nature guiding the way and with such severe consequences, how do we ever end up mismatched?

One argument for how we end up mismatched is that we

don't have the capacity to detect MHC composition using our olfactory ability, especially since we lack the vomeronasal organ, or Jacobson's organ, which is found in the nasal cavity of many animals. This organ is first in line when it comes to olfactory sense and processing. Next time you see your cat smell something and hold its mouth open with upper lips curled and teeth exposed in what is called the flehmen response, you can bet something tweaked its Jacobson's organ.

However, MRI studies have shown that molecules involved in discriminating between "self" and "other" odors, called peptide ligans, activate not just the vomeronasal organ in animals but also parts of the brain.[45] Though humans lack the organ in our noses, we certainly have a brain, and these same molecules, when we smell them, light up the same area in our brain, too. This brings us back to the question then—why do we end up mismatched?

It might have something to do with birth control pills. The irony, or perhaps tragedy, of hormonal birth control is that it interferes with how a woman's nose knows. When women take birth control pills, this natural ability to discriminate between similarity and differences in MHC composition is disrupted, causing women to be more sexually attracted to the odor of males with MHC genes more similar to themselves.[46] Not the best match.

I was discussing this with my friend Stacey, who exclaimed, "That must be why I couldn't stand the smell of my ex-husband!" She went on to explain that when she met her first husband she had been taking birth control pills. Several years into their marriage, after she discontinued the pill, not only was she unable to get pregnant, but she no longer cared for the smell of her husband.

My advice: sniff a potential mate. I personally like the neck. Good for smelling babies and good for smelling men. If you are not on birth control and he (the man, not the baby) passes the sniff test, then that is just one more step toward finding a potentially good mate.

SMART WOMEN ARE SEXY . . . RIGHT?

Although the purpose of this chapter is to discuss attraction based on physical traits, I want to take a moment to discuss one more trait—intelligence. I was curious about whether men find smart women attractive.

We saw earlier in this chapter that within the first tenth of a second of meeting someone we determine not just their physical attractiveness, but also their personality traits, like trustworthiness, general likeability, and competence. Though some might disagree, let's suppose that competence is analogous to intelligence. Does intelligence matter to males when it comes to choosing a mate?

Given my research background, a fair number of my female friends have doctoral degrees. One of them, Roberta, is super smart; she's like ridiculously intelligent. At the same time she is funny, sexy, well rounded, and just an overall great person. I think I'm secretly a little jealous of her. She is a homebody and really wants to find the right guy to settle down with. Roberta's problem? No men seem to want to date her. She's confessed that more than once she has been in mid-conversation with a man when he abruptly stopped talking to her and walked away. What was responsible for this sudden change of interest? In every case, she swears, it was uttering the letters PhD.

Men often say they want a smart woman, but does the evidence really support this claim? Research results confirm what Roberta, and other smart, successful women already suspected. On average, men feel worse about themselves the better their partner scores on an intelligence test.[47] The poorer their female partner performs on an intelligence test, the happier they feel.

This was true even in Nordic countries that are often considered to be more gender equal. This basically means that many men find smart women attractive as long as the woman is not smarter or more successful than they are.

FIRST IMPRESSIONS LAST

Like it or not, our initial attraction to others happens instantly, and it is based largely, if not exclusively, on physical attributes. Facial symmetry, shiny hair, clear complexion, good teeth, and body shape are signals that provide important information about the behavioral and genetic profile of a prospective partner, whether male or female. At the molecular level, we are driven by preferences of which we are not consciously aware. It is entirely possible that like a frigate seabird we are drawn to a particular person because he or she smells just right. On the flip side, whether it is big noses, good hair, or lovely feet that get you going, there is so much variation in what combination of traits individuals find appealing that there is indeed someone for everyone.

So, after some time of questioning my shallowness and quick judgment based on Mr. Flip Flop's not-so-endearing qualities, I gave myself a little slack. Though the dirty tank and unkempt hair were obviously not for me, it could have very well been his scent alone that had me planning my escape from the time he walked in. And, as we all know now, ignoring these hidden evolutionary messages could very well result in a mismatched-MHC-gene nightmare!

3

FALSE ADVERTISING

I recently asked a guy in a bar what was the worst thing that women do. His answer was simple: they lie. When I asked him what they lied about, he replied "Everything!" I know that many women have the same response when asked about men. The irony is that if you take a quick survey among your friends and ask what they think is most important in relationships, many are likely to answer, "Honesty."

We all know that we can't have successful relationships if there is no honesty, no trust, right? Yet there are a thousand little ways in which we lie to a potential mate, never mind our actual mate. Many of these lies can be carried out without our even realizing it, and when we are aware of it, most of the time we consider little lies to be "white lies"—harmless, innocuous untruths designed to attract or appease a mate. This got me wondering: Are humans the only species that lies?

Before entering graduate school to delve into the depths of animal behavior, when people asked me what it was about animals that I loved, I would usually tell them this: "Animals don't lie; with animals, you always know where you stand." My rationale was simple. I had never met a dog that pretended to like me. While this is still true, I have discovered how naïve I was regarding the devious nature of animals. Deception is rampant in the animal kingdom.

One kind of deception we can all relate to is strategic deception. This drama plays out on playgrounds across the globe. Two toddlers are playing on a playground and one of them wants to play with the toy that the other one has. Some children will use crying and screaming to get access to that toy, summoning an adult who comes running over and forces the toy-bearing toddler to give up his or her treasure, reminding him or her that "we all have to share." Immediately a sly little smile comes across the face of the winning toddler as the adult authority hands them the win.

This kind of ploy is very similar to what we see in savannah baboons. These baboons live in fairly large social groups that range anywhere from ten to two hundred individuals. Females often stay in the group they were born into for their entire lives, forming long-lasting social bonds. Social life in these groups is also characterized by a pecking order, or ranking system, among females. If you are a youngster, life can be easier if you are born to a high-ranking mom. Normally when an older, unrelated individual threatens a juvenile, the youngster will scream. That scream will send his or her mother running to the rescue.

Consequently, few low-ranking individuals will mess with a kid from a dominant family. Some juveniles take advantage of the situation and cry wolf when an older individual has possession of a valuable food item, something akin to what the toy on the playground means to a four-year-old child. Of course, once the youngster screams, mom comes and chases off the attacker, and junior picks up the food item that was conveniently dropped.[1]

As we already know, behaviors evolve because they confer some advantage. Therefore, it must be true that under the right conditions the pros outweigh the cons when it comes to deception. Considerable research focuses on why people tell lies; this includes research that explores the underlying psychological defects individuals have that cause them to fib, what childhood traumas we have experienced, and the like. Very little of this research, however, looks at the payoff, or success, achieved by lying. Despite well-known sayings like "Winners never cheat and cheaters never win," by and large we all know that this is not true. Cheaters win all the time. From a biological perspective, if the benefits of lying exceed the consequences of a lie, then it can be advantageous to lie. Period. No need to invoke moral arguments.

When it comes to finding a mate, one of the most common ways we tell little white lies, or in some cases really big ones, especially early on, involves our appearance. We are constantly altering our looks to make ourselves seem more attractive. We spend a tremendous amount of money, time, and energy devoted to this one effort. The Beauty Company issued a report summarizing global market data.[2] Total sales in beauty and personal care worldwide exceeded 400 billion dollars in 2011, with the United States spending thirty-eight billion dollars. Here is the breakdown based on category: hair care (20 percent), facial skin care (27 percent), fragrance (10 percent), makeup (20 percent), and personal care (23 percent). Men are spending more these days, too, with an estimated 27 billion dollars spent annually. You will notice that these estimates do not even include things like weight-loss programs, plastic surgery, or salon costs. Like we saw with sex, if this behavior is adaptive, we should be getting a pretty substantial payoff as a result of all this cost.

It may not be such a huge leap to think that this could have an evolutionary history that predates humans. So when we look to other animals, what do we find? Well, other animals lie about many things, including their looks. The Polyphemus moth is famous for the big "eyes" on its wings that are designed to fool a predator. Or if you are a stick insect, mimicking twigs or leaves trembling in the wind is the way to go. But usually animals do not lie about their appearance to potential mates. Now granted, they may not have a concept of what they look like, but nature didn't evolve to fool a mate, just a predator. Why is that? Is there anything we can learn by looking at why deception (in attracting mates) is curiously missing in the rest of the animal kingdom? Should we be wary of individuals who deceive us about what they really look like? What does deception about appearance tell you, if anything, about a potential mate?

Let's start by taking a look at the ways in which animals are dishonest about what they look like. Then we'll see how we humans deceive each other, explore the costs and benefits of such behavior, and figure out what all this means when we are looking for a partner.

LOOKS CAN BE DECEIVING

As I discovered during my studies and research, from communication to appearance, animals are not as truthful as one might expect. Take the ordinary chicken. When the rooster comes upon some food he gives a call, announcing what he has found. A lot of species give out a food call, especially those that are social, as a way to cooperate when they come upon food. In the case of the rooster, however, his goal is to attract the ladies. To that end, he might exaggerate a little about the quality of the food he has found. Heck, he might even go so far as to call when he hasn't found any food at all! Female hens are always on the lookout for good sources of food, so to that end, the rooster is telling the pretty little hens what they want to hear, even if it isn't true. Of course, as soon as the hen arrives and finds that he has no food, or less than what he advertised, the jig is up.

Fortunately, hens are smarter than many of us. They don't rely on what the rooster has to say to decide whether he's the guy for them. They also don't get overly worked up that he deceived them about such matters. Instead, they look to the size of the comb on top of his head. We'll talk about size and why it matters later, but for now, take my word for it, if it is too small for her tastes, she'll pass him by.[3] Hens intuitively understand what we know as "what someone says is not nearly as important as what someone does," or, in this case, what someone has. Although the male lied to his potential mate, his deception was about his resources, not his appearance.

It was once again back to doing some personal field research on my number one guinea pig—myself. How? Chatting up men in bars, all in the name of science, of course (who ever said science wasn't fun?). I was describing a scenario to two very nice, good-looking twenty-somethings about a guy I had briefly dated. The short version of the story was that the bulk of my dating experience was on the phone, every night for one or two hours. I also explained that he usually called me from work. Now my girlfriends and I had already figured out that this guy was all talk, but I wanted to see what the male interpretation would be. Right away one of them said in awe, "Oh, he was a phone dude. I've never met one of those, but I've heard of them!" He said it as though these kinds of men were the stuff of myths and legends. As for phoning from work, no hesitation, "He's bored." Ouch.

As women—and I will take liberties here and speak generally—we want guys to call us. We love it. There is a saying that men fall in love with their eyes and women fall in love with their ears. Well, I think it's time all of us (men and women) smartened up a bit and started paying attention with all our senses. Now don't get me wrong, it's great when a guy can talk and express himself. And a guy calling is a signal that he is interested. However, if all he is doing is calling, then, no matter what he says, maybe those young men in the bar were right: he's just bored and has nothing better to do.

The point is this: talk really is cheap. That is why the rooster can brag (lie) about what he has (food) without the goods to back it up. This is a very common strategy, and males use it with each other, not just with females. As a woman, I am constantly fascinated when I see human males engaged in a session of one-upping each other verbally.

This verbal one-upping is also seen in the wild. The green frog is a great example of how males in the animal kingdom use the equivalent of "your mama" jibes and insults to compete with each other. Like many frogs, males croak to advertise how great they are. In essence they are saying to females, "Hey, baby, come over here and I will show you a froggin' good time," and to males they are saying, "Don't mess with me 'cause I'm the frog!"

Typically, the larger the male frog, the lower the pitch, or frequency, of his croak. However, some small males are able to lower their croak to make it sound like they are really big, much bigger than they actually are. Basically they are trying to bluff their way out of a confrontation with a larger male. The female frog isn't fooled, however, because, like the hen that assesses the rooster on the size of his comb, she evaluates male frogs based on the actual size of their territory, but other males may be hoodwinked into believing they are dealing with one mighty frog.[4]

There are two things to keep in mind with this example. First, the success rate for these males in tricking females is extremely low. Second, this kind of verbal deception is generally used to convince other males that they are dealing with a bigger competitor, since green frog males fight over a territory. Females, on the other hand, care less about the size of the male frogs voice and more about the size of his territory. Because a smaller male rarely succeeds in securing a larger territory, despite his attempts to bluff, larger green frog males will almost always have larger territories. For females, larger territories signal that a male is of better quality and/or has more resources.

So, ladies, if you are the type who looks for a man with a large . . . territory, then you will want to pay special attention when your date says he will pay for dinner. Make sure he uses cash every now and then. Otherwise, he may be lying about his ability to provide a meal as well as the size of his wallet—meaning that credit cards can very well add up to the human equivalent of the green frog croak, exaggerating territory and resources.

Although I know lots of women who refuse to let their date pay for their dinner, let me tell you a recent story about a friend of mine, Cynthia, since her experience illustrates this point perfectly. A guy at the gym she worked out at had asked her out for a dinner date on more than one occasion. She tried not to turn him down outright because, even though he wasn't at all her type, he seemed like a nice person. But he kept asking. Before her excuses got too transparent, she agreed to a dinner date. She told me later that she figured she'd get a free meal, so why not? Well, when the bill came she reached for her wallet like she was going to cover her share and, as she had hoped, he said, "No, no, I've got it." He pulled out a debit card and the waitress took it away.

Moments later the waitress returned and informed him that his card had been declined. He put that "Oh my gosh, I am so shocked!" look on his face and asked her to try again. According to Cynthia, he didn't look all that embarrassed. When the waitress returned a second time with the same result, her date looked at her and said, "I don't understand. I am so sorry, but would you mind?" Cynthia could not believe her ears. She was completely caught off guard, and she later realized that if she'd had her wits about her she would have asked the waitress for separate checks, paid for her half of the meal, and left him to figure out what to do. As if paying for his meal wasn't enough, he actually said, "Gee, I hope this doesn't affect the way you see me." It did. She said he no longer annoyed her, now she just plain didn't like him.

Now in the case of the rooster, you might be wondering why he doesn't just figure out a way to lie about the size of his comb like some men do about their financial resources. Well, as you can imagine, lying about the size of an actual body part is a bit more difficult. You might think that despite the obvious challenges for the rooster, he could find some way to make his comb appear bigger than it really is. After all, many animals deceive each other about the way they look.

In nature, camouflage is one the most ubiquitous forms of lying. You've got insects that look like sticks or leaves, moths in the various colors and patterns of tree bark, salamanders that resemble plants, snakes that blend in with the desert sand, and, for me maybe the most amazing of them all, the squid that can basically change itself to look like absolutely anything in its surroundings. And many animals that can't change color or shape instead puff up to make themselves look larger and more menacing.

When it comes to appearance, the benefits of many types of camouflage are clear. You live another day, either because you were able to catch a meal or because you avoided becoming one. This inconspicuous lie is almost always used to avoid becoming a meal rather than to attract a mate.

So what about attracting a mate? As we saw in the previous chapter, appearance matters, and there is no getting around it. You are being judged, at least initially, on your physical appearance, and what you look like provides very detailed information to a potential mate. Recall that most judgments we make about a person occur within the first tenth of a second of meeting that individual. So it would seem that if you could successfully alter your appearance to make yourself seem more attractive, there should be a tremendous payoff.

When we review the mating strategies of men and women, we see a major difference between humans and all other animals. Humans of both sexes alter their appearance (sometimes drastically) in an attempt to be more attractive. We are constantly trying to pretend to look like something we are not. The problem that scientists are left to grapple with is whether there is a personal benefit from this behavior.

Take makeup, for example. This is not a modern, contemporary phenomenon. There is a long and even dangerous history of the use of cosmetics in both men and women, which is rooted (whether we want to see it this way or not) in deceiving others.[5]

Today, millions of women and men use a virtual truckload of chemicals and cosmetics to alter the appearance of their face and body. Now some may argue that this is all due to slick advertising campaigns pressuring us to achieve some impossibly high standard of attractiveness. This is certainly a factor. However, long before Madison Avenue was a blip on the radar, makeup was a significant part of human history, albeit, often for different reasons.

Face paint, which is really a form of makeup, has its origins in tribal ceremonial practices during which its purpose is to enhance a story or mimic an entity or a God. In many tribal societies today, this practice is still commonplace. In Egyptian times men and women used makeup for a variety of reasons. Some were practical. Kohl liner around the eyes, for example, was thought to keep away flies, protect the eyes from the sun (much like the eye black football players use today), and ward off infections. Like earlier tribal uses, the design around the eyes also conveniently mimicked the look of the sun god Horus. It's hard to say what came first, Horus or the liner, but the fact remains—judging by the hieroglyphics—that anyone who was anyone had his or her eyes lined with black.

Face paint seems to have been co-opted as a tool for deception around the Middle Ages when pale faces were trendy because they suggested wealth. At that time, lead paints and facial powder made of arsenic were used to convince others you were rich. Since many women died or suffered muscle paralysis due to poisoning, one could argue that the cost far exceeded the benefit. The favoring of pale faces continues today in certain cultures (for example, Japan), where having a white face is considered noble.

In the 1600s things changed, and having a colorful, tan, and radiant face became fashionable because it indicated you were in good health. Rather than actually being healthy, rouge and other heavy makeup were used to cover up a pale face. This at least makes more biological sense. Just as with other animals, a healthy person has good coloring. Think about it: when we see someone who looks sickly we often remark, "You look so pale, are you feeling alright?" This trend continues today, as is evidenced by the popularity among both men and women of naturally or artificially tanning so as to appear "glowing" and "healthy."

It wasn't until the 1920s and the rise of Hollywood films that cosmetics really took off and began being used to set the standard of beauty to attract a mate. We now find ourselves inundated with products, some conveniently called "concealers," that are designed to alter our appearance, often in dramatic ways, concealing our true identity.

Although the standards of beauty can be somewhat different from culture to culture, the irony of makeup trends, at least in wealthier countries, where the bulk of beauty products are sold, is that heavily made up women are not even what men generally find attractive. If you ask them, most men will tell you that they want a real/natural woman and think makeup is "yucky"—yes, that is a direct quote. Yet while some men claim to not understand why women use cosmetics, or say they don't like women who wear cosmetics, their actions often contradict their words. Is there a biological explanation for this?

To solve this puzzle, we have to uncover what makeup accomplishes for women. In strictly scientific terms, glowing skin, flushed cheeks, long lashes, and full pink lips suggest a nubile young woman in good health—the ideal specimen of a human female for the purpose of procreation. Therefore, one could say that men are probably not so much attracted to "more beautiful" as they are to "younger looking" women who would therefore offer the best chance of producing healthy offspring.

But if most guys don't dig the heavily made-up, overdone face, why do so many women overdo it? The most likely explanation is that women are competing with each other in a world where young is beautiful and sexy and older is out. And that's where the barrage of endless commercials comes into play. Though it's hard to say what drove Egyptian royalty to embrace exaggerated black eyeliner, we can pretty much rest assured that in our modern age, it is the cosmetics and beauty industry that is telling women what they need to look like if they want to catch their dream guy and live happily ever after.

Looking at the billions of dollars being pumped into the purchase of beauty products each year, I'd say that it's definitely working. So in the end women are competing against one another with the cosmetics companies telling us how to get the advantage. But let's look at this in another way. What are women really competing for?

Women compete with one another primarily over resources. Where do we get those resources? Historically, from men. While some tribal societies are traditionally matriarchal in nature, for the rest of us, and only in some countries, it is just recently that we have seen the advent of a more egalitarian society. In countries with greater gender equality women are gaining more financial independence, but across most cultures it is still the male who earns more, which translates, for our purposes, to having the bulk (if not all) the resources.

Unlike males, who often compete aggressively, females sometimes do things a little differently. In humans, females compete against each other for mates by trying to outdo other females by enhancing the facial features that males seem to find attractive. Certainly, women spend exorbitant amounts of money trying to look younger, which interestingly is really trying to look more fertile. Does this mean that today's older women are competing with younger "rivals"?

As I already mentioned, we women alter our appearance in a variety of ways, not just with makeup. Many women today opt for hair extensions and hair dye, fake nails, false eyelashes, and deceptive clothing that implies larger or smaller body parts.

Picture this: You are with a guy you really like and things are heating up. You start kissing, slowly then passionately, and he runs his fingers through your hair. Uh-oh, there is a snag, a tear, and suddenly, mid-kiss, off comes your weave and he is now holding your "hair" in his hands. You are embarrassed and start to cry. This loosens the glue holding your eyelashes in place and they tumble down your face. As this unfolding nightmare continues, the look on your date's face, which began as adoration, now progresses through several stages, from shock to confusion to disgust, and finally to horror. You become angry, perhaps indignant, try to grab your weave back (well it did cost a lot of money!), and off pops a nail or two.

Of course, if you go to a local nightclub you only have to wait until closing time to see this kind of scenario unfold. Wait until the lights suddenly go on and you'll probably hear a collective whispered gasp "Aww, damn!" throughout the room as people get a real, bright-light glimpse of who they have been dancing with all night. Normally it is dark in a dance club, and lighting, or lack thereof, can help further camouflage appearance, until of course the lights go on.

Don't be fooled ladies; that quality guy on your wish list will most likely be put off by exaggerated and phony appearances, especially if you're grabbing your hair back and excusing yourself to the restroom. Or worse yet (and I've seen this happen), you are dancing together and the padding you placed in your bra falls out onto the dance floor. Not pretty, and you can count on everyone peeking around to see which woman looks suspiciously lopsided.

We women want to look good, but lying is most likely not the best strategy for attracting a mate. You might be thinking to yourself, "Hey, it's a good strategy, and it works." There's a funny Japanese movie that illustrates this point poignantly—a young woman has a total body makeover in hopes of catching the guy of her dreams. When she finally gets to kiss him, things go drastically wrong. Everywhere he tries to touch her she's worried he's going to feel something strange, so she pushes his hands away. In the end, they both fall asleep on the couch exhausted without having accomplished anything in the romance department.

Just like the rooster who immediately gets attention when he lies about how much food he has, all the chickens—in this case men—might run over to see what'cha got, but if you don't really have the goods to back it up, you'll probably be passed up by many guys. Don't get too smug gentlemen; you are guilty of doing the same things. Take note, the George Hamilton tan, the car salesman hairpiece, and the You Are Not Fooling Anybody Comb Over won't win you quality mates either.

As I've said, men also have a long history with makeup and fake hair that threads its way through the ages: from Egyptians to Romans to Vikings, Victorian nobles, and tribal warriors, right down to today's hyper–beauty conscious men. Men today use concealer, blush, and eyeliner, quickly catching up to women in terms of the money they are willing to spend on personal care and grooming products.

Since a full head of hair is a sign of health in men, it's easy to understand why losing it can be traumatic. It is exactly for this reason that men, throughout the ages, have sometimes gone to great lengths to conceal any signs of emerging baldness. Roman men sometimes lost their hair in an attempt to dye it blond (burned off by the harsh dyes of the time). In Egyptian times baldness was self-inflicted by shaving the head (and the rest of the body) to avoid lice and other creepy crawlies.[6] The solution? Wear a wig. In Victorian England men donned wigs to appear older, more distinguished, and presumably wealthier. There is no lack of concrete examples of this today—hair extensions, hair replacements, hair plugs, hair growth enhancers, and the list goes on.

A discussion of men's hair, or lack thereof, is certainly not complete without mentioning beards. Like my friend Rachel in chapter 2, I am not crazy about facial hair on men, whether it is a mustache, goatee, or a full beard. Maybe, as a behaviorist, it bothers me that facial hair obscures features that give us so much vital information about a potential mate, including his facial expressions.

However, putting personal preferences aside, the natural state for a man is to grow facial hair, so there must be a reason for it. Biologically, it is a secondary sexual characteristic driven by testosterone. Though growing a beard has direct links to various cultural, historical, and climatic factors, there is some evidence indicating that, on average, women consider men with beards to be older, more mature, more confident, and more aggressive, but *not* more attractive.[7] Then why are some modern-day men so intent on growing beards? Perhaps, just as women use cosmetics to enhance their appearance and compete with each other, facial hair in men is more about men competing with other men. Perceiving an angry man with a beard as more aggressive, other men may be less likely to engage in a confrontation with the bearded fellow. As I said, there are certainly cultural differences to take into account, and it may be a cultural preference to shave, but women across cultures, in this day and age, generally seem to rate clean-shaven men, or men with a little bit of a five o'clock shadow, as more attractive.

What about body hair? We certainly have less body hair than our great-ape cousins. One possible explanation for this is that humans evolved to become relatively hairless in response to parasites, you know, fleas, ticks, lice, and so on. Other hairless creatures, like the naked mole rat, live in environments where transmission of such parasites is higher.[8] Group living and being in close proximity to others can increase the probability of catching bugs. Just think of the elementary school class and how quickly head lice outbreaks spread. Could this be why the Egyptians were fastidious about removing body hair, with the exception of the male beard, as previously mentioned? But what about underarm hair? This hair is a reservoir for pheromones, so why do so many women shave it off? In the United States this practice can be traced to a 1915 *Harper's Bazaar* article showing a female model without underarm hair.[9] A campaign was born, and being a hairless "beauty" became a fashion requirement. As skirts were shortened, the hairless legs soon followed, until, by the 1980s, the vast majority of American women practiced hair removal. Men remove body hair, too, waxing or shaving their chests and backs.

Recently there has been a trend to eliminate pubic hair. This is interesting because, once again, pubic hair is full of pheromones. But even this trend may not be unprecedented in our history. Once again, we can look to ancient Egypt, where both males and females removed even this hair. Among the Romans and Greeks, pubic hair on women was not desirable.[10] I have a theory about this practice. My theory is that, like the bare face, bare genitals can't hide problems like parasites or other signs of sexually transmitted diseases. Perhaps the re-emergence of shaving one's pubic hair, in both men and women, is linked to efforts to advertise one's genital health? And rumor has it that the incidence of pubic lice, or crabs, is on the decline, possibly due to this increased removal of pubic hair. I imagine anyone petitioning to list pubic lice on the endangered species list would have to be a purist through and through.

When it comes to deception and appearances, I think that online dating is an experiment all on its own that requires special mention. Online dating is fascinating. I mean, think about it. You look through a virtual catalogue of potential mates and make decisions based on photographs and words, both of which are cheap to produce! We already know people deceive using words. But do people really lie about their appearance through photographs? You bet they do. I decided to be a guinea pig in my own research, and one of the resulting dates ended up inspiring this very chapter.

I had a date for lunch with what seemed like a very attractive computer programmer. I showed up a few minutes beforehand (I like to be punctual) and took a seat at the bar. By 12:15, I assumed I was being stood up. I decided that since it was a nice French bistro and there was a cute guy nearby also sitting at the bar, I would stay and have lunch on my own.

I ordered, and at 12:25 I felt a tap on my shoulder as someone said my name. I turned and was rendered speechless (and sorry, not in the good kind of way) by the man who stood before me. His pictures, if they were even of him, depicted a man at least a decade younger and fifty to sixty pounds lighter. He took a seat next to me just as my salad arrived. I think we were both feeling uncomfortable. I mean, I was obviously more than slightly shocked, and I surely did a poor job of hiding my initial reaction. There was no way he could have felt comfortable knowing that he was nothing like the picture he had sent me. Yup, just awkward!

A survey of people who I know have used online dating services reveals that almost everyone has had at least one similar kind of encounter. Granted, this is based on a small and completely unscientific sample size. And, in fairness, the majority of the people I met did post authentic pictures of themselves, but for those who didn't there is only one thing to say: Seriously people? What are you thinking? It doesn't matter how nice you seemed online, you lied, plain and simple. And it is the kind of lie that is impossible to get away with.

When people make the choice to deceive in such a way, they have to know that it will almost surely be a deal breaker. I decided to put aside biology for a moment to look at it instead from a psychology perspective. By doing this, I managed to have a tad more sympathy for my poor date. He surely *wished* he looked more like his photograph, and he probably hoped that someone would give him a chance even though he started off with a blatant lie. But, in the end, I'm a biologist, and I, like the hen, had run over only to find out that the rooster had, in fact, not told the truth about how great his grub was, and this hen was not impressed. In any case, doctoring photos, in my opinion, falls under the banner of false advertising, and doing so leaves the other person with the feeling of having been deceived—no matter what the latest dating gurus may advise.

There is a more subtle physical deception that I came across, particularly during my online dating experiments. This involved how men reported their height. I am 5 ft. 3.5 in., and yes, I am really going to squeeze every half or quarter inch I can get, even if it's only in the morning. One of the most common deceptions in online dating, other than weight, is height. I found this to be the case with many men who reported that they were substantially taller than they actually were.

Men, listen up. If you are 5 ft. 5 in. and you want to meet a woman who is 5 ft. 3 in., you need to take into account that some of us will wear heels at least three inches high, if not higher. This means that when we meet you for the first time, we will be taller than you unless you are wearing a pair of elevator shoes that can increase your height anywhere from two to four inches. But keep in mind that eventually the shoes will come off, even if the socks stay on. Don't despair; just wait until you read the next chapter, where we take a closer look at size. For now, in this humble dating doctor's opinion, it is much better to just accept how tall you really are.

ANTLERS DON'T LIE

So far I have described to you some of the principal ways in which we lie about our appearance. More importantly, though, I am emphasizing that maybe this approach, if taken too far, is flawed. In essence, these various deceptive strategies ultimately don't work. Why? The simple reason is that animals cannot alter the traits potential mates find attractive. Cardinals, for instance, cannot artificially dye their feathers like we can color our hair. In most animals, then, there are certain features that are immune to lying, and these are referred to as *honest signals*. Advertising for a mate is one area where, for wild animals, anyway, honesty is always the best policy, even if it is by default.

Among birds, signals involved in attracting a mate often include fancy plumage, singing, and dancing. It is typically the male that bears the brunt of this responsibility. In animal-mating systems, the only time females put in that much effort is when they are trying to acquire a collection of males! Otherwise, it is the fellows who need to strut their stuff.

When it comes to birds, the female is focusing on particular features, for example the color of his feathers, the color of his beak, and/or the size of his tail. As we saw, the reason for this focus is that the way these features look is often *directly* linked to the health of a male. That is why the male with the brightest colored feathers will usually win the girl. The male cannot cheat. The only way he gets brightly colored feathers is if he has a good diet and few parasites. A female bird is relying on these signals to inform her mate choice.

Imagine if every Tom, Dick, and Harry cardinal out there could just apply some red paint to his body. How on earth would a female be able to determine whether he is a high-quality male? She simply wouldn't be able to, and the consequences for her, and for him, would likely be severe. Their offspring might not survive because he would not be fit enough to provide food for everyone.

Here's a question: Have you ever wondered why there aren't elk, deer, or reindeer, for that matter, running around with hollow antlers? If you are an animal behaviorist, these are the things you think about. Seriously. With a cup of coffee in one hand and a far-away look, these are the thoughts running through my head.

It takes a vast amount of energy to grow antlers. Among the cervids, or hoofed mammals, usually only the males have antlers, and these antlers are made of bone. Yes, bone. The process is the same for most species. The antlers start from permanent bony pedicles, or stalks of tissue, located on the male's head. Initially, antlers are covered with a type of "velvet" tissue that is infused with blood vessels. Once the antlers are finished growing, the velvet "dies," the males rub their antlers on trees and shrubs to get the velvet off, and the antlers mineralize into bone.[11]

Sounds like a piece of cake right? Wrong. Males are adding almost 30 percent of their body weight in bone and need to increase their food intake significantly to do so. Not to mention that just growing antlers can require energy intakes up to five times the normal metabolic rate needed to sustain the male. That's like having to consume five times as much food as your body actually requires, which many of us do, even though we have no antlers to show for it!

Not only are antlers expensive to produce, but once you've grown them, you have to lug them around with you everywhere you go—at least for a while—which takes a tremendous amount of effort. Then you just drop them off somewhere in the forest and start all over again. All this cost begs the question: Why not just cheat and grow hollow antlers? It also begs another question: Why shoot these guys only for their antlers when you can just pick them up in the forest? I'm just saying. . . .

If you see an animal sporting a weapon, from the antlers on elk to the gigantic horns of the dung beetle, you can bet there is some serious dueling going on between the males. So if there ever was a mutant elk whose antlers were hollow, you can bet he didn't last long. Two possibilities would have befallen our imaginary male. Since most males would prefer to bluff their way through a dispute rather than actually come to blows, and they are realistic about their abilities (unlike most humans), the first possibility is that our imaginary male bows out of every potential fight. In doing so, he would not have the chance to mate. Why? Because females watch these fights very carefully, and the bottom line is that losers don't mate.

I know, it's harsh, but it's the harsh reality so close to Darwin's heart. Larger males grow larger antlers, and larger antlers typically win in a fight. If a male doesn't win a fight and subsequently mate, this hypothetical rare mutation will never be passed on, thus no hollow-antlered males. The second outcome is that the male is delusional (not uncommon) and thinks that his antlers are just fine. Well, the first time he gets into a dispute with another male with proper antlers, he's toast. He might even die. Either way, he doesn't get the girl.

Though nature comes down hard on those in the wild who lie about their own attractiveness, in the spirit of full disclosure, I must report that to snag a mate sometimes animals do lie about their appearance—sort of. There are a few *rare* instances where lying through your proverbial teeth about your appearance to get a mate does go on in the wild, albeit not exactly in the direct boy-meets-girl human way.

Remember our green treefrog from earlier? Some might argue that the frog is deceiving other frogs. But, if you remember, in that example the male just croaked about his size—his actual size wasn't deceptive. Plus, as we agreed earlier, the deception is less about attracting a mate than it is about scaring other male frogs away from entering his territory, although there are some instances where this deception works to draw a potential mate close enough to check the male out. We can also look to the mighty fiddler crab to see another kind of subtle deception.

Fiddler crabs belong to one of the few groups of organisms that have been detected to cheat during male-male aggressive interactions and successfully fool females. When it comes to deceit and courtship, fiddler crabs are joined by snapping shrimp and hermit crabs. You might already have picked up on a similarity here. They are all crustaceans.

There are a lot of species of fiddler crab, but they share some similar features. One can find them along beaches or intertidal areas around salt marshes. The easiest way to recognize them is by the males, who have one large fiddle-shaped claw. Sometimes these crabs look like they are waving at you with their one big claw. Indeed, the males are waving, not really at you, but at other males and at females. This waving is designed to signal to both males and females how big and strong they are.

You could probably guess by now that the size of the claw matters. Males will fight for territories, and larger males with larger claws are typically stronger and will win fights against smaller males with smaller claws. Are you seeing a trend here? Females again pay attention to the size, but here females use the length of the claw as a measure of attractiveness.

Herein lies the opportunity for cheating. Like all crustaceans, fiddler crabs shed their exoskeleton, or outer shell, as they grow. Sometimes males lose this larger claw, the one they wave around. When this happens males can grow a new claw, not unlike a lizard growing a new tail. The difference is that this new, regenerated claw is identical in size (length) but not mass to the original. It is lighter, and most importantly, it is weaker than the original claw.[12] So, while the male is somewhat weaker, he waves around his claw, successfully bluffing both males and females alike into thinking he is still the stronger version of his old self.

Another example of successful visual deception involved in mating happens in orchids. Since examples of wild animals deceiving mates are far and few between and orchids do something pretty amazing, it is worth mentioning them here. Plus, you may never look at your orchid the same way again.

When it comes to pollination, most animals are enticed—bribed, if you will—by the flower to pay it a visit. The flower provides sweet nectar (sugar) to the pollinator in exchange for stopping by. Instead of saying, “I love you very much,” flowering plants say "I trade you very much." Orchids, however, play a different game altogether. Nearly one-third of all species of orchid are liars. They lie about food or they lie about sex. A common strategy among some orchids is to entice a pollinator by deceiving male insects into believing they are landing on a sexy female insect and about to get lucky. Think bug version of a vagina. The depth of their deception is profound. Some orchids even mimic the odor of a female insect by producing compounds that smell like pheromones given off by a female bee. Eau de Insect, if you will. The males in their enthusiasm pounce on the orchid, ready to mate, only to discover that it is no sweet honey of a bee waiting for them, just a plain old orchid. Fortunately, there is no walk of shame for such males and inevitably they will repeat their performance on another orchid.[13]

This fake copulation results in insects picking up and depositing the pollen from one flower to another. Many orchid species have even evolved to trick only specific species of insects, thus ensuring their pollen will be spread only to flowers of their own species.

The prize for mating deception has to go to the lightning bug, or firefly. However, even though the deception in this example has to do with mating rituals, there's a little twist at the end. Fireflies are like the magical creature from everyone's childhood. They flit about on a summer's evening, little flashes of light, or bioluminescence, everywhere you look. What you may not know about fireflies is that all that flashing going on is not for our amusement or pleasure. Instead it is all about sex. That's right. They are flying around sending out a signal via light that they are there and available.

Each flash is unique depending on the species. In other words, each species flashes at a particular speed, and each flash lasts for a certain amount of time. As you might imagine, different species of firefly may not look all that different, so this is how they recognize each other. When a female sees a flash, she responds, and this initiates a kind of light show between the male and the female reminiscent of kids sending coded messages by flashlight clicks.

Under normal circumstances the male approaches the female, and let's just say, sparks fly. Unfortunately for males, some females of a different species lie just for the sadistic purpose of getting a man. They copy the flashing pattern of the kind of girl he's looking for. Then, when he lands on her, she eats him. Whoops! The winner here is clearly the deceiving predatory firefly.[14]

What does all this mean? The laws operating in nature are in place for a reason, and they simply do not support deception when it comes to mating. Which begs the obvious question—what's going on with us? Why are humans the only animal species that goes to such great lengths to deceive when it comes to attracting a mate? Do we want to be fooled, or are we all on the lookout for the imposter among us?

Often, though not always, we can look at appearance and spot a fake a mile away. This observation skill is there for a reason. Having babies is serious business, and the vast majority of us, whether we are conscious of it or not, want what we see to be what we get. Indeed, just like our wild cousins we need what we see to be what we get.

TAILS AND TITS

If I told you that there were scientists out there attaching fake tails to birds and fish, you would probably think I was joking. One could argue that we behavioral scientists are a weird bunch, but the research we do is useful to all of us. So, for those of you out there still convinced that artificial coloring and surgical implants are really the best ways to snag a mate, this section is for you.

Large breasts, large muscles, enormous glutes; what if we are "programmed" to notice these things or even have a built-in preference for them? A trip to any adult store will suggest that human females have preferences for unusual sizes *and* shapes as well! Are there any animals out there that have a similar preoccupation with conspicuous, outrageously large physical characteristics such as these? C'mon, by now you should know the answer!

One particularly enchanting example is the long-tailed widowbird, or *sakabula* in Zulu, that makes its home on the grassland plateaus of Kenya. Females look rather like a house sparrow, while the males have one of the most extreme sexual ornaments found in that genus of birds. What is it? If you are thinking back to the peacock, you got it. Half or more of their tail feathers are a half-meter long (or approximately one-and-a-half feet)!

Males perform graceful displays during which they pop up and flutter like butterflies over the grassland, their long tails trailing behind them. By carrying around such a long tail, male widows are so conspicuous that they can be seen over half a mile away, putting the male at grave risk for being attacked by predators such as raptors. To balance out this huge cost, there must be a payoff, right? There is. Just like the peahens, females dig widowbirds with longer tails. They find them super sexy. When male widowbirds have their tails cut, their sex appeal drops significantly. On the other hand, if you artificially enhance a male's tail, look out![15]

What is this female preoccupation with long tails, you ask? As with other bird species we have seen, it seems that long-tailed males are superior in that they have better diets and fewer parasites, therefore endowing them with the surplus energy to fly around with all that extra tail. However, this does not explain how the association between long tails and sexiness came about in the first place. Since most of the evidence suggests that female preference is responsible for some of the weird and wacky appearances found in males of many species, this is an interesting question.

Here is where all those experiments with fake tails come in. If you take a species that does not normally have a long tail (therefore there is no link between male quality and tail length) and you artificially add a longer tail or some other ornament, will females go gaga for those enhanced males? Incredibly, the answer is yes! And it doesn't even have to be a tail.

Let's take a look at the collared flycatcher. Of four species of flycatcher, the collared flycatcher is the smallest, a modest little bird. The male has a white collar and a large white forehead patch. Some clever researchers decided to see what would happen if they gave some males a red stripe on their otherwise white forehead patch, like a Mohawk. Female collared flycatchers had never seen this (or so we think), and since it was a brand-new trait, there would be no connection to male quality.

Typically when collared flycatchers mate, males that get there early and acquire a breeding territory (by competing against other males) do better. What the researchers found was that among males that arrived early, there was no real difference between the red-striped males and normal males. However, if you got to the party late, having a red stripe on your forehead apparently helped you out considerably. Not to mention that once a female went red, she rarely looked back.[16] Meaning those females that had experience with these artificially colored males seemed to have developed a preference for them and never wanted to mate with the now "ordinary" males. Too bad for the females; that is, unless researchers plan on forever painting red stripes on the foreheads of some lucky males.

Is this why some human females prefer males with tattoos or conspicuously colored hair? Maybe we aren't so different from some of these creatures after all. The real difference here is that, without the help of scientists to attach fake tails or painting stripes, there is no opportunity for wild animals to artificially enhance their appearance in this manner.

Like people, birds are funny. When it comes to studying birds, one of the ways in which we identify individuals in nature is by placing little colored rings around their legs in different patterns. Early on it was observed that the particular color of the band could seriously alter the course of an individual bird's life.

The consequences for zebra finches are the best studied. These beautiful little birds live in a variety of habitats, primarily feeding on seeds. They are very social, and the males are enthusiastic singers, with fathers passing their song on to their sons. Males that received red bands typically won more contests against males that happened to have green bands, particularly when it came to fights over food. If you win more fights over food, you become fatter and sexier to females.[17]

At first, it was thought that females just liked red because it mimicked the red bill of the males, but if that were true then there should have been no differences in success among males wearing green, blue, or orange bands. However, those unlucky enough to get green fared the worst. Maybe females just really hate green? Not likely.

In zebra finches heavier males (in this case males that get red bands) are in better condition and can afford to spend more of their time singing and displaying. The more a male sings, the more attractive he is to females. Who doesn't like being serenaded? Of course, no one has figured out why red-banded males win more fights against males that get bands of other colors. Maybe, like in people and rhesus monkeys, the color red signals power, dominance, and enhanced competitive performance, causing all the other birds to be intimidated.[18]

The work with zebra finches does not end with color bands. What about a random blue feather attached to the forehead of a male? Surely this would make a male look strange, different, and unattractive, right? Not a chance. Once again, males with a blue feather attached to their forehead were considered ultra-sexy by the females. As if that wasn't enough, female zebra finches that were raised by fathers with a blue feather stuck to their head had daddy issues later on and preferred males with a blue feather coming out of their forehead. Males, on the other hand, did not imprint on the blue feather and did not spend their mating days desperately seeking a girl finch with a blue feather sticking out of her forehead.[19]

So far most of the examples have been with birds, so you might be thinking that birds are very strange indeed. It so happens that the fascination with exaggerated traits is not unique to birds (or humans). Fish can be included, too. Specifically guppies, wild guppies—yes, they do exist in the wild. Male guppies have a lot of variation in their appearance. They have spots on their bodies, and these spots can (1) appear in different places and (2) come in a wide array of color combinations, including blue, green, red, black, yellow, and orange. When given the choice between males that had a common color pattern versus males that had unique color patterns, females went for the unique males.[20]

There are many ways to look at these examples, but for our purposes, the take-home point is that, in addition to the traits we already discussed, novel, unique, useless, and downright outrageous traits can be appealing. The appeal may be due to exposure at an early age (blue feather) or because it relates somehow to mate quality (red band vs. green band; long tail), or it can simply be because an individual is different (color pattern).

Any one of these can imply that an individual possessing these traits is of good or better quality. What doesn't add up, if you place all these examples in an evolutionary framework, is that by artificially enhancing traits, we are in essence undermining the power of natural selection. Sorry Darwin. The poor little zebra-finch girl grows up looking for that perfect blue-feathered mate. But since her father had been artificially "enhanced," that trait will not be able to be passed down and has nothing to do with the actual health of her parent or her favored mate.

This effectively means that human intervention managed to create a false message. We can easily see that faking in nature has the potential to upset the entire balance of a species and therefore evolution itself! This brings us to probably the most important question of all. Are we doing this to ourselves? Have we thrown Darwin's theories out the window with the bath water?

One human physical trait subject to mass enhancement is breast size. Some have been made to be as large as watermelons! Now, if you say, "Men prefer large breasts," this is not necessarily an accurate statement. Lots of men do, for sure, but not as many will actually say they prefer fake ones to real ones. Let's assume there is a preference that is common and that over time (evolutionary time, that is) this led to human females having naturally exaggerated breast size compared to other mammals (which is actually true). Is there any advantage gained by having this preference? There is no evidence linking larger breasts with better quality milk production or lactation ability. This means that breast size is not a good predictor of how well a woman can feed a man's offspring.

So what gives? One idea is that the reason men prefer larger breasts has nothing to do with size per se, but similarity of size between the pair. As you may recall, the sameness, or symmetry, between the left and right side of any feature reflects the genetic, developmental, and overall health of an individual. The symmetry between the left and the right breast is a good predictor of the reproductive ability of a woman.[21] This means that men could be unwittingly assessing the health and fertility of a woman. Maybe not so surprisingly, as breasts become larger, it is easier to detect size differences between the pair. This could also possibly explain the aversion some males have to fake breasts. They are being deprived of important information.

These days, it is not simply fake body parts, but a whole suite of artificial treatments and fillers that hinder our natural ability to assess the quality of a mate, at least based on physical appearance. Some treatments even reduce our ability to detect facial expressions or patterns, which is crucial for successful social interactions. It is no wonder we are all having such difficulty!

PRIMPING, PREENING, AND ROCK STARS

At this point you might be wondering what is natural when it comes to our appearance? Isn't there anything natural we can do to enhance our attractiveness without upsetting the Darwinian order of things? Well, of course. We can be the best version of ourselves naturally by highlighting features that already exist. Animals do this all the time. Let's take a look at a few examples that illustrate how animals accomplish this.

Take the splendid fairywren, or superb fairywren, whichever you prefer. When the male is of breeding age and condition, his plumage is a beautiful turquoise, such that it would match the color of most splendid waters in the Caribbean. He has patches of turquoise on the top of his head (crown), cheeks, and mantle, while his breast is a deep blue hue. If I sound like I am in love, it is only because I have a serious fondness for blue/turquoise.

So how is it possible to enhance his already brilliant blue? When it comes to bringing flowers to court a girl, human males don't have the market cornered. The male splendid fairywren collects yellow petals and uses them to woo the females. In this case though, the petal isn't for the girl. It's for him. He holds it in his beak while he displays by puffing out his cheek feathers, lowering his tail, and twisting back and forth so the female can see how beautiful he is—the fairywren version of a flamenco dancer.

Though a few males choose pink flowers, yellow is more complementary to blue, so perhaps that is why most males choose yellow. It makes me wonder whether blue-eyed men buy more yellow roses.

Although the pretty petal may complement and highlight what a brilliant blue the male really is, the female also bases her choice on when males acquire their plumage; the earlier, the better.[22] Why? Only older, fitter males can afford to molt into a better breeding plumage early on in the breeding season and court the female over a longer period of time. The take-home message is that a hopeful male can bring all the flowers he wants, but he will not achieve the desired effect if he is dull to start with.

In my humble opinion, the rock star of enhancing his appearance is the bowerbird. There are many species of bowerbird, but my favorite is the satin bowerbird—probably because his plumage is an iridescent blue. In general, bowerbirds are notorious for the fact that males build nuptial bowers, a structure designed to attract and entice females to mate.

There are two main categories of bowers: avenues and maypoles. The avenue bower starts with a foundation of twigs. The male then inserts two walls that arch up and sometimes inwards. In contrast, maypole builders start with a growing sapling and build a cone around its base, sometimes extending nine feet. Regardless of the type of bower, males take the building task very seriously, collecting and meticulously placing items so that their bower is just perfect.

Once the structure is in place, they set about decorating it. That's right, they decorate. Surprise, surprise, the iridescent blue satin bowerbird goes gaga for anything blue. And I mean anything. In this, they are not so discriminating. Blue plastic, blue flowers, blue ribbon; anything blue they can get their beaks on. Don't be fooled; it is vicious out there in the bowerbird world. Males will routinely steal blue items from other males, and they may even outright stomp and crush another male's bower. Once a male has completed his bower, the females come through and inspect it. If a female likes his bower, he then must dance for her.[23] What girl wouldn't want to be a bowerbird chick?

What can we learn from fairywrens and bowerbirds? Accessorize. Know your best features and use things that draw attention to them. The bowerbird male successfully does this by surrounding himself with things that bring out the best in him.

Most of us do this already, though not all of us draw attention to our best features. We wear hats, scarves, jewelry, and other types of glitz and glamour. How to accentuate your features successfully, whether you are a man or a woman, has been the topic of many popular television shows. There is nothing deceptive about choosing to wear items or colors that best complement the features you already have. Rather than changing ourselves to fit the idea of what we think someone else might be attracted to, we can utilize these tools (that are all at our disposal) and become the best possible versions of ourselves. The outcome is well worth it, not only from a biological standpoint, but also from a totally human one, since in the end you will find the person who chooses you for you. And that, according to this guinea pig, is the basis for any relationship that has the potential to stand the test of time.

4

SORRY GUYS—SIZE MATTERS

There I was on a Friday night enjoying a beer at a local watering hole before heading out to the clubs for a little dancing. Two women came up next to me, squeezing in to order drinks. We all smiled, and then suddenly this voice behind them said, "Here, let me get your drinks ladies. I want to buy them for you." I made eye contact again with the women, who gave me an eye roll, conveying a silent message among women that can only be described as annoyance.

I looked behind me to see the source of their irritation and found a stocky, blond-haired man whose upper-body thickness obscured his neck. He continued, saying, "C'mon, ladies, I got it. Really!" They looked at me and shrugged, the universal message here being, "What the hell? Free drinks."

He bought them their drinks, and now that they were sufficiently obligated to talk to him, I heard him ask this question: "Okay, if you had your choice between a tall, good-looking man who had no job and a short, good-looking man, such as myself, who had a good job and made a lot of money, who would you choose?"

I confess I could not resist. I turned around, smiled, and said, "Wait, can I answer that?" The women didn't seem to mind, so I said, "They would marry the short one with the good job and sleep with the tall one on the side." Both women nodded, while he just sputtered, "Who the hell are you?" Who, indeed?

Most of the time when we talk about size and men it is a reference to penis size. But as this situation illustrates, maybe things aren't so straightforward. What about body size, testes size, or the size of a man's brain? His bank account? His heart?

Perhaps my answer to the question posed by this guy was biased, since in the past I have been particularly attracted to tall, muscular men. I had never given this preference much thought, until recently. Now that I was evaluating my dating life from this new perspective, I wondered—where does this preference for larger body size came from? To take it a step further, how much does size, regardless of the currency, really matter?

BIG BENEFITS

Not so long ago a colleague of mine suggested that perhaps some women ("like you") prefer taller, muscular men because a larger male is a better protector. Though it certainly wasn't the first time I had heard this theory, for some reason, it made me stop for a minute to think about my preferences. I realized that I certainly felt "safer" with such men. I know this isn't the case for all women. I have a friend who thinks she's invincible and makes fun of me for what she calls "insecurities." Insecurities or not, for some reason, I automatically assume that in a dangerous situation a larger, well-built man will just handle it, presumably so I don't have to.

In the wild, we often see a significant body size difference between males and females. Do these bigger males actually protect females better? Is male body size akin to the peacock's ridiculously long ornate tail? Meaning, is the reason men are on average 10 percent taller than females purely because females find larger males super sexy? To find this out, we first have to see if being big has advantages in the wild.

ARE YOU LOOKIN' AT ME?

In many cases, males are bigger than females because males fight with each other. What do they fight over? Girls, of course. Usually when there is combat between males, the larger male will win the fight. Sometimes this aggression is subtle, as in the case of one group of butterflies, the longwings.

You may be wondering what on earth male butterfly hostility looks like. I mean, it's not like we see butterflies flying around beating each other up. Yet, believe it or not, they do just that. Male combat in this group consists of a considerable amount of pushing and shoving. The wrestling gets really intense when there is a female nearby, and the bigger male usually wins.[1] Doesn't that just make you want to go out and watch butterflies?

Since we are on the topic of fighting without limbs, snakes present another unusual form of combat. If you have ever seen two snakes, their bodies raised in the air and entwined with one another, you might be inclined to think it was a romantic interlude. Chances are, you would be wrong. It's apt to be two males who initially raise their bodies partially off the ground, signaling to each other that they are ready to begin. At that point, it is game on, and all bets are off. There is pushing. There is shoving. And their bodies entwine. The goal, much like wrestling matches in schools all over the world, is to flip, pin, and subdue the opponent. Again, the larger male usually wins.

Not all males fight fair though, and some are not above taking a cheap shot. Like the snake equivalent of a kick to the nuts, vipers are willing to fight dirty. I doubt the Renaissance poet John Lyly was thinking of the Australian blacksnake, a viper species, when he wrote, "The rules of fair play do not apply in love and war." Nevertheless, one study describes a situation in which a male was already mating with a female and a second male started trying to push his way in, literally.[2] Seems it is really difficult to get a male to stop having sex with a female once he starts (you think?), and the mating male ignored the intruder even after being pushed into the water.

A third male showed up on the scene and then a fourth! Male #3 decided this was just too much, and he vacated the scene. Male #4 launched an all-out attack on Male #2, biting him in the neck. Heads were raised, and their hoods spread, then the two snakes collapsed around each other, writhing and wrestling. Head pushing ensued, and both snakes tumbled down the bank and into the water, where they continued fighting. Male #4 apparently won the contest, and Male #2 made a hasty exit.

Now, the copulating pair was oblivious to all this drama, continuing about their business in ordinary snake fashion; that is, until Male #4, possibly invigorated from his win, slithered on over to drape himself over the female and bite the mating male in the middle of his body. Props to the couple—after forty-five minutes of continuous mating, this interloper deterred neither, even though he continued biting the male. At some point all three tumbled into the water, and unbeknownst to Male #4, the happy couple quickly drifted downstream and he was left "scratching" his head trying to figure out where they went.

Contrary to the scene depicted above, rarely do males just start scrapping out of nowhere. Oftentimes, the confrontation is very calculated; before the first blow is dealt, males size each other up, frequently choosing not to fight at all. That is because fighting, whichever way you do it, takes so much energy and carries so much risk of injury or death, that you definitely want to pick on someone closer to your own size.

Much like boxing, where opponents are put into weight-matched categories, males of drastically different sizes usually don't fight each other. However, when two males of a similar size encounter each other, the competitors may circle each other, carefully assessing their opponent.

Male eastern grey kangaroos engage in ritualized fighting very similar to boxing. When not in combat mode, males are pretty tolerant of each other, hanging out and feeding close together. When not fighting, male kangaroo etiquette for an approach involves touching noses or sniffing each other's head and shoulders. What is kangaroo etiquette for a fight? Float like a butterfly, sting like a bee, hop like a kangaroo, and punch, punch, punch! If the blows from your paws don't cut it, use your tail as a prop and kick. As if winning wasn't enough, the eastern grey kangaroo male then rubs it in by doing a victory dance, grabbing a bunch of grass and throwing it at his own chest.[3]

Like kangaroos, do human males assess each other? Certainly they do. I have made several interesting "field" observations while out with a man—especially if he happens to be particularly big or muscular. First, men of smaller stature puff themselves up when larger men pass by. Second, these smaller men greet the more muscular man (and only the man, never me), by emitting a gruff, cutoff version of hello, how ya doin', or even just a chin lift as they pass. Third, and arguably more fascinating, the man I might be with—if he's larger—slightly slumps his shoulders, raises his hand up, gives a small smile, and says, "Hey, how's it going?" It is almost as if the larger man is saying, "No worries, I'm not looking to kick your ass, though we both know I can."

These behaviors bring back memories of my dearly loved Great Dane. He, like most Great Danes, was truly a gentle giant. For those brave little male dogs who dared to come up to us in the park—all puffed up and barking obscenities—he would simply lower himself down until he was lying on his stomach with his head on the ground. This shrinking body language was often enough to convince the little guy to come over and sniff him. Many a time, he would be able to make friends and soon would be running around and playing (carefully) with his new little buddy.

Alright, ladies. A great place to see this play out on a very human scale is in sports bars or at pool halls. Try it and see what happens. Specifically, position the male you're studying where other men have to pass by—the corridor to the bathrooms can work well. If your guy is a little on the smaller side you may see him puff up a little. I would also predict that while smaller guys will usually be more concentrated on your bigger date, a larger, more muscular man will have no reservation about making eye contact with you. If your guy is on the bigger side of things, you may observe him unconsciously reassure smaller men. Either way, it's interesting to observe, and doing so may show you a side of your man you've never noticed before.

In fact, if you aren't around, there is a strong probability that they may not even bother doing it. How hard males fight largely depends on who is watching. It's almost as if they say to each other, "Well, there are no women here for us to impress or win, so why bother killing ourselves? Beer anyone?"

This seems to be the approach used by the leaf-footed cactus bug. As its name suggests, it has "feet" that look like leaves, and males try to take control of a cactus that females need for laying their eggs. Sometimes males fight before a female arrives. Other times, the males begin fighting only when a female arrives on the scene. The interesting thing here is that when a female isn't present, male size has only a small influence on who wins the territory (cactus). However, when females are present, the big guys pull out the big guns and size predicts who wins.[4]

I guess bigger, more "muscular" leaf-footed cactus bugs need a bit of incentive to put in the effort. And they certainly aren't alone. Scientists who conducted a study on the male-male aggression of grey treefrogs hypothesized that a lack of females being present may be why body size was not a good predictor of fighting success in staged fights between male frogs. There simply was nothing to spur them on.[5]

Speaking of being spurred on. . . . Sometimes size may not matter as much as experience. The idea, then, is that believing you can win, because you won once before, might be enough to do the trick. Officially it is called the "winner effect."[6] From cockroaches to humans, winning once may sometimes give you a leg up on the competition, but it isn't likely to completely overcome the effect of size. For instance, cricket males that had previously won a contest were more likely to win even if they were smaller, as long as there was no physical contact.[7] In other words, before things escalated to a full-blown grappling war, the winner effect came into play. However, as soon as males are able to assess each other's strength through physical contact, the larger male will win. Winning on your home turf enhances this effect even more.

And what about us humans? For men it is enough simply to believe that they performed better than a male competitor. Yet another reason, ladies, to provide positive feedback to your man about his gallant performance.

With all this fighting, and all these large males winning, does that mean that bigger males get more girls? In animals, pretty much. In the kangaroo example, the heavier males are usually more dominant and sire more offspring. So, right after his grass throwing victory dance, he can be seen heading off into the sunset with the girl.

In these examples males are not protecting females at all, but females are attracted to larger males because a male who wins a fight is generally perceived to be the better quality male. Perception is not always reality though, and bigger males don't necessarily always pass on better genes. In brown trout, even though females prefer older and bigger males, this does not translate to offspring that are more likely to survive.[8]

Though nature provides us with ample examples of when bigger is better, there are also plenty of instances when it isn't. And thinking you are the winner, no matter what your size, goes a long way. So, keep a positive attitude, and no despairing. Bigger is simply not always better.

CAN I OFFER SOME PROTECTION, MA'AM?

When males are charged with the responsibility of protecting females, the benefits of size are more clear-cut, and the bottom line here is that a larger male can usually do a better job. What exactly do females need to be protected from? Primarily two things: predators and other males. Male crickets tend to hang around after they have mated. Although some thought the reason they do this is so the male can make sure no other guy comes and displaces his precious sperm, the reality is that male field crickets are trained in the art of chivalry. When males and females are alone, they are both at equal risk of being killed by a predator. But when a male and female are together and there is a predator attacking, the male lets the female hide and get into the safety of the burrow first. She rewards him by preferentially mating with him (rather than with many other males) more frequently, and, as a result, he fathers more of her little cricket babies.[9]

Aside from the obvious benefit to females of being directly protected from predators, there is another advantage to having a male around. In a given day, there are time constraints. We can all relate to that. There are only so many hours to get things done. Animals spend most of their time doing a few things: sleeping, eating, moving about, working on house repairs, and watching for predators. The more time you spend doing one thing, the less time you have available for other activities. Therefore, if a male is around keeping an eye out for predators, a female can devote more time to other things, like eating.

Harlequin novels may be about romance, but male harlequin ducks know how to protect their women, allowing them to rest and eat so they can be at their peak. Whether the female is in her fertile period or not, her partner is four times more vigilant with her than he is without her, particularly when the couple is feeding close to shore, a very dangerous area for these ducks.[10]

Because male birds often help raise the kids, they tend to

hang around. Some male birds happily feed their female partner while she sits on the eggs all day, but not grey-headed juncos and red-faced warblers. Females not only have to incubate the eggs, but they mostly have to get their own food, too. When males were experimentally removed, females spent more time alert to predators, less time foraging at their full capacity, and even less time sitting on their eggs.[11] So even though the male didn't help with the chores, it still paid to have him close by.

The behavior of these males is very similar to what many human males do when they take a woman out to dinner. Often, the male partner takes up the position that allows him the best vantage point to identify a potential threat. Haven't you noticed? Perhaps they are just protecting themselves (they want to be aware of a threat immediately), but the byproduct is that females can eat in peace, blissfully unaware of any potential problems and secure in the knowledge that should anything come up, their male companion will handle it.

Two of my male friends, in particular, do this, but to be honest, I think it is more a trained response in them rather than a conscious desire to protect their dates. Both are in law enforcement, and both served in the military. Whenever I go out for drinks or dinner with either of them, it is second nature for me to check that I have chosen the riskier position (for example, a seat that positions my back toward the door). Given that I have complete confidence that they have my back, literally, I am very okay with this.

Nowadays, the majority of us are not under threat from natural predators, but there are certainly some man-made threats, such as cars and other hazards on the street or sidewalk. Men who take the role of protector seriously always walk on the side closest to cars, buffering females from potentially getting hit. It is considered good manners. This doesn't necessarily mean, though, that when faced with a natural predator the same man won't run to save himself.

A friend of mine, Lizbeth, relayed an incident to me that captured this perfectly. She was spending time with a new beau who was tall, muscular, and conscientious. He always opened doors for her, took up position on the dangerous side when they were walking together, and generally made it a point to look out for her.

One evening, while out night fishing, they were walking back to his car talking about the potential wildlife out in the forest at night. She jokingly asked, "If an animal came out of the woods would you protect me?" With complete seriousness he said, "Not a chance. If an animal comes out of those bushes it is every man for himself." At that very moment there was a rustling in the bushes (most likely a squirrel), and true to his word he started to run. Though he only ran a few paces before realizing embarrassingly that it was nothing, it was pretty clear that he really would run first and check on her later.

Whether or not a man walks on the dangerous side of the road or takes one for the team, most men bristle at the thought of another man harming or harassing their girl. Male harassment of females is not just a human problem, as unreceptive females in many species have to deal with persistent males. Although we will discuss this in more detail later, a few examples are pertinent here. For instance, don't let the permanent toothy smile of the dolphin fool you. In Shark Bay, Australia, male bottlenose dolphins of one group will work together to kidnap females from other groups and hold them hostage for over a month![12] Female bottlenose dolphins are promiscuous, and not all females mind being herded away by gangs of males, but for the ones that resist, the aggressive male bullying, which includes hitting and biting, sometimes results in injuries.

Bottlenose dolphins aren't the only kidnappers out there. Hamadryas baboons live in clans in which the social group is subdivided into smaller subgroups, also referred to as bands. These bands usually contain a single adult male, his harem of females, and occasionally a younger male that is a follower of the leader.

Males from other bands within and outside the clan will try to abduct females. Why risk kidnapping a female? It would appear that the Stockholm syndrome is not exclusively a human response. If kidnapping hamadryas males are successful in keeping the female away from her male long enough, the absconded female will bond with her captor. And it doesn't take long.

The tale of Mick and Julie, two hamadryas baboons, illustrates the lengths to which a male will go to get his girl back.[13] Unlike other older males, young Mick only had one girl, Julie. One day he appeared alone; Julie was nowhere to be found. He searched for her, and when he spotted her in the midst of another band, he ran to her and the two embraced. A male from this foreign band reached out to touch Julie, and Mick promptly attacked him. Unfortunately, because Julie had been gone for a whole night, she was not so keen to return to Mick. The three fought for almost an hour, with everyone sustaining injuries, until the other male gave up. Julie was so severely wounded that Mick had to drag her back. Poor Julie. In other abductions, in which females are only gone a short period of time, they eagerly escape when their male comes to retrieve them.

In contrast to the kidnapping schemes of the hamadryas, some male savannah baboons try a different approach. Rather than kidnap a female, they offer to protect her from other males. Thus, females form long-term friendships with a few particular males in order to escape broader male aggression. These males protect their female friends and their infants from others in the group. This is definitely a friendship with benefits, as females are more likely to mate with their friends.

I don't know about other women, but I certainly benefit from the friendship of a few key males. I have four male friends who at any time will unequivocally assist me in dealing with any type of harassment.

I faced a situation that is not that uncommon for a single female living alone. In this instance, a maintenance worker for the apartment complex I lived in came to do repairs. While there, alone with me in my apartment, the male worker made a sexually explicit remark that left me feeling extremely unsafe in my own home.

Later, when I pondered that this man potentially had access and keys to my apartment, I panicked. I sprang into action, calling these trusted male friends to ask what they thought I should do. One offered to harm the offending male (I declined the offer), another offered to come and stay with me so that this person would think a man lived with me, and all helped me go through the steps to handle the situation so that I could make certain I was safe. As a strong, independent, confident female, I have no problem letting a male protect me, or open a door for me for that matter.

More often than not, females do need males to protect them from other troublesome males that try to mate with them. In most cases, it is a top priority for a male not to lose his female to another male, so protecting her is a win-win for both of them. Stallions are key protectors of their mares. Females of many equine species, including zebras, stay close to a particular male, especially when they are lactating. When nursing their young, the female Grévy's zebra is harassed four times more often than nonlactating females.[14]

Why in the world would males harass females that are already caring for offspring? In the case of Grévy's zebras, a female that is harassed loses about ninety-six minutes worth of feeding time. More importantly, the female moves around pretty fast to try to get away from the harassing male, and this puts the young foal at risk for becoming separated from mom. If separated, the foal can die. If her foal dies, a female is more likely to reproduce again quickly, giving the harassing male a chance to mate with her and father her next foal. This is a form of indirect infanticide via harassment. Though it seems pretty drastic, in other species it is more straightforward. Males will outright kill infants. They do this because the female will often immediately go into estrus following the death of her offspring.

Gorillas are the reason I became an animal behaviorist. I am madly in love with them, silverbacks in particular, and I am mesmerized in their presence. They are critically endangered, and I shudder to think that this world could very well lose them. Gorillas are the largest, and in my opinion, the most magnificent primate. Silverbacks are enormous, weighing in somewhere between three hundred and five hundred pounds, and they stand anywhere from 5 ft. 6 in. to 5 ft. 11 in. Females, in contrast, are about half the size. Why the big difference? Silverbacks face little sexual competition from other males in the group, but the silverback must protect his females and offspring from outsider males. As you probably already can guess, body size, once again, is a better predictor of a male's success. But why are outside males so dangerous?

The silverback usually lives with a group of unrelated females and offspring of varying ages, and he has many responsibilities. He leads the family unit to food sources, makes decisions about where to go next, settles disputes among his ladies, and protects the females and young. A successful silverback can remain the leader of a group for as many as fifteen years.[15] Because of this long tenure, young male and female gorillas usually leave the group when they reach maturity. This is because young males can't compete with dad and young females can't mate with dad.

It is pretty obvious that females don't have too much trouble joining another group, but what about males? A silverback will not tolerate an outside male for understandable reasons. So how is a bachelor male supposed to start his own group? Well, unless he can kill the silverback (which rarely happens), stay in his original group with his father and wait for him to die (which could take a very long time), or take over a group that has lost its silverback, the best strategy is to kill an infant. If an outside male is successful in killing a female's offspring, she will more than likely leave and mate with the male that killed her baby. Why? Because her male failed to defend her. Among gorillas, if you want to mate and successfully raise your young, it pays to be big.

What about us? Is body size really that important to human females when we are evaluating a male? When I really thought about it, I realized that I cared more about muscles than height. Males are cross-culturally on average 10 percent taller than females, and although there is some research that suggests taller men have more children, it doesn't seem that height is what is most important to women.[16] Rather, it is muscularity. Women, on average, it seems, strongly prefer muscular men. Specifically that glorious v-shape, where chest breadth is greater than the waist size . . . ahh! So it is not height, but thickness that does it for us. And not just in body size, guys. This also applies to penis size.

JUST SAY IT: PENIS SIZE

There is considerable variation out there in penis size, and little of it has any direct relationship to body size. If we think about my beloved gorilla male, he is enormous, but his penis is, well, tiny. Possibly two inches when erect. If you really want to insult a guy, tell him he is hung like a gorilla. One can't look to body size as a good predictor of penis size. That also goes for hands, feet, or other body features. Many a woman has learned this the hard way, muttering to herself, or more likely to her girlfriends, "But his hands were so big!" Yeah, we talk about it. Sorry guys.

In contrast, turtles have it going on. Relative to body size they have colossal penises. This begs the question, why have such a large penis when you are pretty small? Is there a benefit to having a larger penis? Naturally. One possible benefit is that you can show your penis off to other males in an aggressive display designed to beat out the competition. Yes, penis displays are pretty common, especially in turtles.[17] While it's unlikely that human males are whipping it out to show it off to each other, many a man has confessed to taking a sneak peek while at the urinal.

What is more common among human males, though, is displaying penises to females. I am not talking about the flashers who want to show their penises to the world. Talking to my friend Roberta about this chapter, she said, "Wait, I have the perfect story:" After a night out at a comedy show, she went to a local pub near her house, where she knew the bartender. There were four men hanging together, and when she sat down and ordered a beer one of them came up to her and asked her why she wasn't there with her boyfriend. Classic.

She informed him that she had no boyfriend, and he pointed to the best-looking guy among his friends and said, "My good friend Magic over there doesn't have a girlfriend. Maybe you two should talk." Of course his real name wasn't Magic, but no doubt he believed his penis was delightful. When the clothes came off later in the evening, this European stranger pointed to his penis and said, with a heavy French accent, "Look at my penis! Isn't eeet fantastic?" Kind of like a turtle. Perhaps many men think this, but few say it out loud.

But I digress. Back to large penises and their potential benefits. By and large females of most species, including humans, mate with multiple males simultaneously. In other words, promiscuity is not just for the boys.

Of course, males are all about making sure that their sperm is the one that fertilizes the female's eggs. The bigger your penis, the more easily you may be able to displace sperm from other males. Let's return to the turtles and ask whether they are well endowed just so they can show their penises off. Probably not. There are several possible reasons for their massive penises. One idea is a practical one. In girl turtles the cloacae (turtle vagina) is not at the base of her body under the shell; rather, it is farther up, near her midsection. With all that shell getting in the way, the male's penis has to be pretty long just to reach it.

One might find the eight-foot-long penis of the blue whale impressive, but you have to remember that relative to body size (blue whales are about one hundred feet long), it just doesn't stack up to the competition. Instead, it is a tiny crustacean that is second in line for the prize for the longest penis. A penis always wants to reach its goal, and if the girl is far away and you are a barnacle that can't move, well, you just might need a really long penis to get to her. It would be even more helpful if you could let your penis do the walking. Acorn barnacle penises are designed to do just that. If you are an acorn barnacle living in crowded conditions you don't need a penis that can stretch as far. But if fellow barnacles are far away, their stretchy penises can reach out twenty-five millimeters, even though they may average only five millimeters in size.[18] Sometimes the size of your penis may just have to do with reaching a potential mate.

A second possibility is that size *and* shape together can provide an advantage. Not only are turtle penises big; they are also complex-looking structures with ridges and grooves. It is thought that these ridges, grooves, and folds assist either in getting their sperm closer to the goal or in getting rid of any sperm left in there by another male.

The prize for the longest penis probably needs to go to the Argentinean lake duck, as the male's penis measures almost seventeen inches even though that is almost equal to the size of the rest of his body.[19] Given that only 3 percent of birds still have a penis at all, we should congratulate the Argentinean lake duck for having a penis almost the size of an ostrich. It is also shaped like a corkscrew, matching the internal shape of the female. With such a long, complicated penis, you could lasso a female who might be trying to get away from you. Not an uncommon occurrence in ducks!

As we will see, when it comes to the simple unadorned look of the human penis, we should consider ourselves lucky. Though if you go to any adult sex shop it would seem that some women like penises with a variety of different frills on them. The days of Tupperware parties may be long gone, but in some cases, they have been replaced with the sex-toy party.

A few years ago I received a gift from a very close friend. It came from one of these parties—an absolutely enormous purple dildo with rubber barbs and a little rubber attachment that looked like a rat mid-shaft. With so many buttons on this thing, I was surprised it didn't light up and spin! I should clarify that this was a gag gift to embarrass me at my birthday party (thanks a lot, by the way!), but it has relevance here. In many animals, when the penis itself is not large, another option is to have one adorned with physical structures like spines, barbs, and hooks, almost like weaponry. Ouch.

For males the added structures serve two purposes: they (1) improve fertilization success, and (2) discourage females from mating with other males by, well, injuring them. In the lake duck mentioned above, the really special feature of the penis is not just size, but that it is shaped like a corkscrew with a soft bristle top and hard spines at the base. Since most ducks are promiscuous, it is thought that the spines and bristles act like a brush for removing the sperm deposited by other males, thereby improving chances at fertilization.

Another creature, the seed beetle (or bean weevil), illustrates this principle as well. In this species the male has a penis heavily ornamented with giant spines. The bigger the spines on his penis, the more eggs he fertilizes, while the female unfortunately gets scarred (literally) by mating with a male with large penis spines. However, it does not prevent her from mating with multiple males, so in the end it is only the size of his spines, not the degree of scarring, that determines his success.[20] This once again brings us back to promiscuity by females as the driving force behind some of the penis-shape patterns we see.

Where do humans fall on the spectrum of penis size? In contrast to our primate cousins, human males, in general, have proportionally larger penises relative to body size. The human penis is roughly twice as long and twice as thick as that of chimpanzees. This may indicate a preference by females for larger penis size in humans, but, like body size, research shows that women prefer girth to length. The average length of a man's penis is approximately 5.5 inches. However, if you have a long penis but no girth you will probably not be favored over a man with a shorter, thicker penis. Are we predisposed to like a thicker penis? Maybe. Is there a benefit to having a thicker penis? There just might be. While human males don't have spines on their penises, that doesn't mean that the shape doesn't matter. The overall shape of the penis didn't just come about randomly.

Biology is such a wonderful area of study because there is truly something in it for everyone. So if you have a burning desire to, say, investigate why the human penis looks like it does, you can have at it in the terribly noble pursuit of biological science. It has been speculated that the shape of the human penis is adapted for the same purpose as the turtle's and the lake duck's—to displace the sperm of other males.[21] This means that female sexual promiscuity is a necessary ingredient to this formula, otherwise there would be no selective pressure driving this adaptation.

What are some of the special features of the human penis? The mushroom capped tip, or glans, for starters. Unique when compared to our closest relatives, there are similarities in appearance with the glans found in other species, like turtles. Another feature? The coronal ridge. This is the rim around the underside of the glans, possibly analogous to the seminal ridge found, once again, in turtles.

It turns out that adult sex shops are not just beneficial to their pleasure-seeking customers, but to science as well. Through a series of experiments using latex vaginas and artificial penises it was discovered that the thrusting action during sex helps displace fluid and that the coronal ridge acts like a spoon. The deeper the thrust (longer penis) and the wider the coronal ridge (girth), the more fluid was removed. The bulk of the job though fell to the coronal ridge, removing two times as much as an artificial penis that lacked a ridge—over 85 percent more. Not bad!

Another interesting finding: if your male partner has recently accused you of cheating, he may thrust deeper and faster next time you have sex as an unconscious attempt to get rid of any other sperm that might have been left there by another man. The same goes for couples that have been separated.[22] When all is said and done, a longer penis may be able to thrust deeper, thus depositing sperm closer to its target, but you only have to have a penis long enough to get 75 percent of the way there. A thicker penis (at least at the coronal ridge), on the other hand, can make all the difference.

SILENT COMPETITION: SPERM WARFARE

At this point, once again making these sweeping statements based purely on statistics, if you are tall, muscular, and have a longer, thick penis, you are in pretty good shape. Or if you are short, muscular, and have an average but thick penis, you are still good. But there is something more. Just when you thought we were done with looking at features of physical size, we cannot ignore the testicles. This brings us once again to female infidelity and competition among males, especially sperm competition.

In some species, testes size is inversely proportional to penis size. What sets males with big testicles apart is sperm with a superior competitive edge. Their sperm are faster and longer (bigger) and can get to the egg first. A positively ingenious way to make your sperm longer, thereby increasing the chance of fertilizing the "golden" egg, is to create a sperm train. In many murid rodents, a group with over seven hundred species, including the European wood mouse, the sperm have a little structure called an apical hook. The male's sperm, once inside the female, link together, creating a chain.[23] This chain of sperm can move faster and with more force than any single sperm could. So a single male's sperm will organize and cooperate to outcompete another male's sperm. In species with proportionally larger testes (an indication of the intensity of sperm competition), the apical hook is longer, presumably to better facilitate linking up of the sperm.

It is well documented in mammals that larger testes size is associated with not just bigger sperm, but also greater sperm production, faster moving sperm, and more frequent mating. These are all indications that males are competing with each other at the level of sperm. When it comes to testes size, humans fall somewhere between gorillas and the highly promiscuous chimpanzees. Like the penis, there is little evidence that human male testes size is correlated to body size. To put it in the basest of layman's terms (no pun intended), it means that being tall doesn't mean you will have big balls.

Although the jury is still out on the effect of infidelity by females on human testes size, men with larger testes could have an advantage when it comes to fertilization success simply because they ejaculate more sperm.[24] What doesn't seem to have been explored yet is whether females have a preference for testes of a particular size. Therefore, I speculate, relying once again on my smallish and scientifically humble sample size, and suggest that given a choice of small versus large testicles, females consciously or subconsciously prefer larger ones. I sense an interesting study that needs to be conducted.

DID NAPOLEON HAVE A COMPLEX?

Maybe Napoleon had a complex because he was short and thin and had a short, thin penis and small testicles. It doesn't seem that height should really have been such an issue, especially since he was 5 ft. 7 in. tall (average for his time). But, then again, that fact doesn't seem to stop smaller-statured males who seem determined to have an issue with height. The primary reason may have to do with interactions with other males, not females. As we saw above, smaller males tend to lose conflicts when paired with larger males.

What, then, is a smaller-sized male to do? One strategy is to be hyperaggressive. Dr. Alfred Adler, founder of individualized psychotherapy, probably wasn't thinking of desert goby fish when he inadvertently coined the phrase "Napoleon complex" by using poor Napoleon as an example of how short men compensate for their feelings of inferiority with excessive aggression, but he could have been.

Gobies in general are pretty remarkable. Several species can tolerate extreme temperatures, while others handle low levels of oxygen, broad ranges of pH, and even high levels of salinity despite being freshwater fish. As if that wasn't enough, they can change color to blend into their environment.

While overall these are small fish, the smaller the desert goby male, the quicker he is to attack an intruder. These smaller males immediately and fiercely attack before a larger male has a chance to assess the size of the territory or resources and figure out that he could beat the smaller male in a contest.[25] It is a risky strategy, but not uncommon. If a small desert goby fish had sat on Adler's couch he would have been told that striving for power and dominance had become pathological, since he was clearly excessively aggressive.

Is there evidence for this in human males? Surprisingly, there have been few studies closely examining this issue. It seems as though convention has it that short men use exaggerated aggression as one strategy to compensate, and we all just go along with it. How much of this is true? We can all probably identify with the stereotype, and most of us know that one short guy who is overly aggressive, generally obnoxious, unnecessarily loud, and a tad too focused on the gym. Sometimes I have even encountered groups of them out together, almost like a pack. Seriously. Have you seen this? Especially in nightclubs, they move through like a group of angry warthogs.

Although I have already mentioned my penchant for dating tall, muscular men, I decided to purposely go outside my comfort zone and date some men of shorter stature and less muscular build. Of course, all of this was done in the name of research (and because I still hadn't found Mr. Right). Maybe it was time to see if Mr. Right was hiding in a different body type. A few of these smaller-statured men met the stereotype, overcompensating for height (and maybe one or two other things) by being overly aggressive. Since I find this behavior distasteful for many reasons, it led to the rapid dissolution of my interactions with them.

One fellow, though, stood out in particular. Paul. He was 5 ft. 3 in., and I probably should not have worn my high-heeled shoes. However, his height was not the biggest problem. Though he was gentle, considerate, and interesting, it seemed like both he and I were feeling uncomfortable. For me, however, it wasn't his height that was the biggest problem. After a few dates I concluded I simply could not date a man who weighed only a few pounds more than me, and who I could probably beat in an arm-wrestling match.

As for Paul, he was clearly uncomfortable as well. It was pretty clear when he kept asking if I owned any shoes without heels. His obvious lack of confidence was not helping his cause either. I almost wanted to tell him about the "winner effect" to help him out a little. While overt aggression is a turn-off for humans, so is lack of confidence. Though it is a fine line to walk, a healthy dose of confidence and feeling good in your own skin and bones, no matter how long or short they happen to be, will go a long way in attracting a mate.

Culturally, there is a belief that taller is better, and taller men do earn more money, but so do more attractive men (and women for that matter). This makes disentangling height problematic. The general consensus is that if you feel good about your body, you feel good about yourself, and subsequently you perform better. While it is shown on the whole that shorter-than-average men seem to feel worse about themselves, they still perform just as well as, if not better than, their taller counterparts.[26] This suggests, once again, that height is not a predictor of success in human males.

Dr. Adler would probably suggest that shorter men overcompensate by being overachievers, which would fit into his theory of them having a "Napoleon complex" regardless of their success or failure. But did Napoleon really have a complex? Was he really just a little tyrant pissed off at the world because he was born short and had visions of grandeur? Or was he a simply brilliant leader who happened to be of average height for males at that time?

BRAINS VERSUS BRAWN

Maybe Napoleon got the short end of the stick when it came to Adler's use of him to exemplify an inferiority complex. Napoleon is thought to be one of the greatest military leaders of all time, and, as I said, his height was average for men in his time, as it still is today.

It is unlikely that his height was responsible for his temperament, since he is self-described as a domineering brat, even as a child. It was his father who was able to secure an education for Napoleon and his brother at the expense of the King of France. It may have just been coincidence that led Napoleon to be educated for the army while his brother was educated for the church.

Regardless, Napoleon excelled and went on to change battlefield tactics, territorial governance, and the very course of history itself. One could argue that, with the introduction of the Napoleonic Code, he was far ahead of America when it came to civil liberties. As emperor, he implemented many laws that promoted fair government, including ones that forbade special treatment based solely on the status of one's parents, respected and protected religious freedoms, and disallowed nepotism in acquiring governmental employment.[27]

Even if Napoleon did have a complex, he clearly chose intellectual rather than physical prowess to achieve status. This may have been what made him so attractive. After all, power is an aphrodisiac. Just ask female Japanese macaque monkeys, whose frequency of orgasm goes up with more dominant males.[28]

And using your brain to gain power is a great option. In Jane Goodall's historic study on the chimpanzees of Gombe, there was one male in particular who seemed to embody this approach, Freud. Unlike other alpha males who might rule with an iron fist, when Freud took over as alpha male he was more inclined to use peaceful tactics. He was smaller than his younger brother Frodo, but Freud used clever ways to enhance his displays. He often groomed other males, and he was less aggressive, rarely beating up the girls, though he was by no means a pushover.

Intelligence has clear reproductive advantages, particularly if it affords you a competitive edge over others. One of the main benefits may be that you are smart enough to use different strategies depending on the situation. You know when it pays to be aggressive and you know when it pays to try something different. Male brown-headed cowbirds with greater social intelligence are better equipped to exercise behavioral control and regulate how they deal with different competitive situations.[29] The end result is that these males are more successful, producing more offspring.

You may remember the satin bowerbirds, where males go

bananas for anything blue, which they use to decorate their bowers in an attempt to lure females. Not only are they masterful home decorators, but some males are smarter than others. Bowerbird males love blue, but they strongly dislike red, at least near their bower. So distasteful is the color red that males tested for their cognitive problem-solving abilities did not require food as the great motivator, red tiles super glued on their bower was all it took. I know, it's kind of mean. Can you imagine the distress? As if that wasn't enough, the tiles were covered with a clear plastic container, and males were tested to see how long it took them to figure out that they had to remove the container to get to the tiles. Overall, males that were better at solving the problem of red tiles also had better mating success.[30]

Maybe the smarter you are the more resources you can acquire. This matters to females because females, in many species, are also sizing up males for their resource potential—for themselves and for their offspring. Therefore, a male's problem-solving abilities could also make him better at meeting the diverse needs of a female.

Lastly, smarter males may have better genes that lead to smarter offspring, if intelligence is heritable. The smarter kids should be more successful and more likely to survive. In species like the grizzly bear, where the male's job is finished with impregnating the female, it's usually the biggest and not necessarily the smartest male that "wins the prize." We can make some sort of correlation to human "short-term" partners. People who have affairs are much more likely to choose partners that are highly symmetrical—making that choice primarily based on physical attributes.

This would seem to make sense. For example, if you're looking for the mother for your future children, though you might still be head over all googly-eyed for the fully bosomed blond with a tiny nose and a perfect petite chin, you will almost assuredly have lots more to assess before deciding if she's the right woman to have your children.

It's the same for women, for whom the pool boy, with his ripped abs and sculpted shoulders, is a real-life temptation—c'mon, we've all seen the movies, and the woman is probably not too worried about how intelligent he might be. A fling with the pool boy may be fun, but, like men, when women are looking for a long-term partner, they are looking for something deeper than the exterior package. Luckily for men without the pool-boy physique, women tend to find smart and funny men very, very attractive. Like in animals, being intelligent often predicts access to resources (for example, money), holding a position of power (that is, dominance), and having better social standing.

Women care about these things a lot. When given a tiny budget to "purchase" traits they found important in a mate, effectively forcing women to prioritize, intelligence was deemed a requirement while other features were considered an indulgence. Creativity also ranks high on the list, as does humor. However, studies show that if a woman makes her own money, male intelligence becomes less important and physical appearance once again takes precedence.[31] I guess that would explain rich women and their boy toys!

BEING A LOVER, NOT A FIGHTER

At this point, you might be inclined to think that mate selection is all about fighting and competing, whether males use their bodies or their brains. With all being fair in love and war, some battles are won in the bedroom. Good news for all—some males excel by being attentive, expert lovers.

Previously, I mentioned how Japanese female macaques climaxed more frequently when they were paired with a dominant male. It seems as though staying power may be another aphrodisiac. How skilled a male is at physically stimulating a female contributes heavily to female . . . ahem . . . satisfaction. In macaques this apparently means how long a male lasts and how many mounts and thrusts are involved. You may be wondering just how long a male has to last and what exactly a female orgasm looks like. Out of sixty-eight copulations observed, none exceeded a minute and a half, and female macaques turn back, clutch the male, and occasionally nip him when they are climaxing. Love bite? Sometimes the females even prefer new and unfamiliar males to the resident dominant male. We see similarities in humans, where females are more likely to experience orgasm with their lover than their main partner, but more on that later.

Recent research suggests that better quality macaque males are kinder, gentler, and engage in more positive behaviors designed to retain a partner and keep her faithful.[32] Yes, keep *her* faithful. Lower quality males, on the other hand, are more apt to threaten and rely on manipulation, intimidation, and possessiveness to keep a female.

Another study of macaques shows just how valuable positive interactions can be. Male mating success ultimately doesn't depend only on dominance rank (which is often correlated with body size) or even on staying power. Rather, males who become good friends with females, groom them often, and have positive social interactions with them do even better than those other males. There is definitely something to be said then for being nice.

5

ON BEING A CHOOSY FEMALE

When I was growing up in South Florida, Palm Beach was the place of the rich and sometimes famous. Estée Lauder had a home there, as did the Kennedys. This was long before South Beach became the "it" spot. What I distinctly remember was the advertisements for seminars on how to land a rich man. Included were things like how to dress, how to act, and most importantly, where to go to find such men.

As a behavioral ecologist I can see the value of one aspect of that approach, even if I fundamentally disagree with the others. Since females of most species are indeed interested in resources, seeking males with sufficient resources makes complete sense. As for the other aspects, the reality is that for a whole slew of species—most of them, actually—it is the males that must look a certain way, act a certain way, and go searching for females. Yet in humans, although studies show that the table seems to be turning slightly (for example, metrosexuals), for the most part, it is still the female who does the lion's share of the primping and preening so that she can impress members of the opposite sex, even those males who do not have abundant resources! Have we women been sold a cultural bill of goods?

Let's look at what is going on out there with animals. As I said, under most circumstances, males pursue and impress females. When it is the other way around, it is because females are trying to acquire a collection of males. But I am jumping ahead of myself. Back to males chasing females. If males are chasing females, what are females doing?

In the field of behavioral ecology we have a saying. When

it comes to mating strategies, "males follow the distribution of females, and females follow the distribution of resources." Here is the tricky part—what do females consider to be resources? If you start with the obvious, it's resources that meet the physical needs of the female and/or her offspring. This might include things like food, water, shelter, and so on.

Within a given territory, a female may have access to all the food she and her babies will need. Another resource the female may be looking for is the genetic material provided by a male. Or, in simpler terms, females want males for their sperm. Like territories, not all sperm is created equal. Finally, females may be seeking a male that is a good father. The caveat with all of these resources is that they are not mutually exclusive. In many case, females are interested in acquiring all of them.

BEING CHOOSY AND EYEING UP THE MALES

The first thing females are doing, then, is assessing males for their ability to provide them with resources, whichever resource or combination of resources they need. The second thing they are doing is deciding whether to mate with a male that meets those criteria.

Evaluating criteria involves deciding whether what you see is really what you will get. This brings to mind another one of my memorable dates. Jack had an interesting story. He explained that he was involved in litigation that promised to leave him well endowed financially. He told me this early in our communication, prior to our first meeting. With some experience under my belt, I was suspicious. Nevertheless, I proceeded to arrange a meet-up at a well-established restaurant chain known for its moderate prices.

At my eatery suggestion, his first response concerned how much he disliked the chain because it was expensive. A not-so-subtle sign of a bluff? He was a few minutes late, but he had called, explaining that his cousin's car had broken down and that he'd had to make other arrangements. Translation: *I don't have a car.*

My first thought upon seeing him was that he was physically my type, so the gene criterion got a check. I also noticed that he was a flashy dresser, with gold chains around his neck, a big, expensive-looking watch, shoes that probably pushed the $200 mark. Not exactly my preferred type, but considering the information I already had gleaned about him, I was now thoroughly intrigued. Okay, so maybe his car was in the shop?

As we settled in for our meal, I discovered I was the only one eating, as he "wasn't hungry." Translation: *You are paying for your own meal.* Given that, I ordered whatever I wanted, including a big glass of wine to wash down the bad feeling I was having. In spite of that, the conversation flowed fairly easily.

He asked what I did, so after telling him, I took that as my cue to ask what he did for work. His response was, "I am making myself available to take calls from my lawyers whenever they need to reach me." Followed by, "I also am staying with my mom to help her after her kidney transplant (last year). She works part-time, so I help her around the house." Translation: *I don't have a job and I depend on my mother*. That was strikes three, four, and five.

Just like females in every species, women are wired to try to obtain the best resources possible. And we know that throughout history—right down to today—with few exceptions, males have controlled most of the resources. Since women need them and men have them, women subconsciously or consciously assess men's valuable resources, or, in Jack's case, the lack thereof.

In the absence of unambiguous cues, how do females assess males? In wild animals, it is almost always the male who is primping and preening, desperate to impress a female and convince her to pick him. The female uses this behavior to determine whether he's the guy for her. We will look at male strategies in detail, but for now suffice it to say that males frequently cajole, serenade, dance, fight, display, or downright harass females.

We know that females are looking for an ideal combination of resources, but that combination is not the same for every species, or even for individual women. Let's first talk about genetic resources. For female prairie chickens, the male provides neither food nor parental care, so her primary concern is the quality of his genes. In the animal version of ladies' night at a dance club, males congregate in an area where they know females will be found. They then carve out a territory called a booming ground. And on this booming ground they keep other males away while they dance, display, and vocalize.

I used to see the exact same thing at a club in downtown Raleigh, where a local group called the B-Boys dominated the space. I considered myself fortunate to be the honorary female member of the group, which resulted in several perks, ranging from protection to always having a dance partner.

But back to prairie chickens. While the boys are strutting their stuff, the female prairie chickens have their pick as they saunter through, evaluating each male. How can the females tell which male has good genes? Here, a male's lasting power, for . . . ahem . . . dancing, may be the key. Males have enormous stamina in this regard. They don't do it for a day; they don't do it for a week; and they don't even do it for a month. Instead, males display constantly for two months![1]

The swashbuckling dance involves the male prairie chicken inflating his breast sac and fanning out his tail feather—not unlike the human version of the chicken dance that is probably being performed at some wedding somewhere at this very moment. Speaking of which, dancing matters in humans, too. We evaluate men based on their dancing ability, and the chicken dance just won't cut it guys. If you think about it, nightclubs are really like the human equivalent of a prairie chicken's booming ground.

There are two primary thoughts on why we look at how

men move on the dance floor. First, it may be a way for us to visually assess their physical fitness, athletic abilities, competitiveness, and stamina. A male who can move and shake better than others is deemed to be of higher quality.[2] This certainly could be said for the aforementioned B-Boys. From 10 p.m. to 2 a.m., their endurance was unmatched by other males, and many, many females took notice.

On the other hand, there is some evidence to suggest that how a man dances is linked to his sexual prowess rather than his physical fitness. Although females look at eleven different components of a man's dance moves, in the end there are three that count the most: (1) how he moves his neck, (2) what's happening with his trunk, and (3) something about the right knee really gets us going. It is not the constituent parts, but rather how they all work in concert that we find so mesmerizing.

Speed matters, too. Picture a dancing man who is making big moves that incorporate his head, his neck, and his upper body. He is swaying and twisting to the beat and simultaneously bending his right knee. Okay, maybe it *is* a little like the prairie chicken dance! The only movement that seems to predict actual physical strength is forward, backward, and side bending of a man's upper body. The question, then, is why we fancy all the other moves, plus the eight others not even mentioned. Perhaps they convey a sense of body coordination? After all, men that were rated as better dancers were also more symmetrical. There is that darn symmetry rearing its head once again.

What if it's more than just good genes a woman is after? Maybe she wants real estate, too. We hear it all the time when one is talking about starting a business—location, location, location. The general rule that cuts across species is that females need a high-quality territory for their offspring to thrive.

European pied flycatchers exemplify this principle. Pied flycatchers are primarily insectivores adept at catching bugs while they are flying around. The males are good-looking birds, with black-and-white patches distinguishing them from females. Not only is their plumage flamboyant, they also sing their little hearts out, hitting complex notes that they believe will win over a girl of their choice. How glamorous and flashy a male is, though, means very little to females. The only thing they care about in the end is the quality of the nest site. Females gauge the worth of the territory based on how safe it is and how close it is to a preferred food source.[3] Not so different from human females. We also like a safe neighborhood that is close to shopping.

Speaking of shopping, for some females, receiving gifts from a male is the preferred way to size a guy up. Yes, we are still talking about animals here. These gifts can include food, rare resources, or even attention. When it comes to food, males feed females directly in close to 30 percent of bird species. This is also common with crickets, butterflies, and spiders. Perhaps it's why we like going out to eat so much.

But in other species, where gift giving by males is not a prominent feature, females will still trade sexual favors for food. It really is a profession as old as time. Like humans, female chimpanzees don't usually do the hunting, but meat is a prized possession that supplants the dominance hierarchy that normally characterizes chimpanzee communities. The male who makes the kill owns the meat. Other males, including the alpha male, will beg for even the tiniest piece of meat.

A benevolent and, perhaps, wise male will share his meat with others. This is because, normally, even though the alpha male tries to dominate a female when she is receptive, female chimpanzees mate with multiple males in the community. And the generosity has some nice perks—females show favoritism toward a male who shares meat with them. When a male shares his meat with a female, she is twice as likely to mate with him later.[4] If you are a bonobo, you have a lot of sex, period. But if a female fancies what a male is eating, she doesn't wait to mate with him later in exchange. No, she will saunter right up to him and present her glorious bits right in front of his face. Who could resist? Not bonobo males, that's for sure. While he gets distracted and begins mating with her, she not so subtly takes the food he was eating for herself.

All this food for sex, and meat in particular, doesn't seem so different from hunter-gatherer societies and possibly modern ones as well. On more than one occasion I have received a better cut of meat, meat at a reduced price, or one or two extra pieces thrown in from the males behind the counter at a local supermarket I frequent.

Of course, I have no intention of mating with them, and I don't think they are deliberately trying to bribe me, but from a scientific perspective I am absolutely fascinated by this behavior. And I am not the only one. Several women I know have received a "gift" of meat from the males working the butcher counter.

Food is not the only gift females are interested in receiving in exchange for sexual favors. Adélie penguin females use rare stones to build nests, and their fidelity to their mates is challenged every time they encounter a male who offers them a stone.[5] Okay, *challenged* is a bit of an understatement. Despite being in committed relationships, females find the stones irresistible, and they readily cheat on their partners.

We humans like rocks, too, specifically diamonds. Where did this practice originate? It wasn't until the fifteenth century that diamonds became popular. The first recorded incident of a man presenting his fiancée with a diamond as an engagement ring was in 1477, when the Archduke Maximilian of Austria gave Mary of Burgundy a ring with diamonds set in the shape of an *M*.[6]

Until the mid-1800s, when mining improvements made them more accessible, diamonds were rare, and basic understanding of simple economic principles makes it fairly easy to conclude that they were also very expensive. How does this make the diamond significant for human courtship? Just as the peacock needs extra resources in order to develop and carry around his long, flashy, and frivolous tail, so, too, does a man need extra resources in order to afford something as nonfunctional as diamonds, or jewelry in general.

However, just as some spiders try to trick their potential partners with an empty gift, perhaps the cubic zirconium helps men pretend they actually have these resources. Some women seem to have learned this lesson from female spiders that verify their gift contains food by judging its weight. These women head over to the jeweler to verify that the stone is real. The bottom line is that diamonds and other precious stones signal a male's resources.

But there is a catch. There always is, ladies. For those women who found out their ring didn't have a real diamond, it's even worse than you thought. Sorry. The higher the value a man places on his potential bride, the more he is willing to spend on a ring. Therefore, whether it is a diamond ring, property transfers, or dowries, women who receive gifts know, or at least sense, that they are valued. This means that there is actually a biological reason why we girls like gifts so much. And, no, vacuum cleaners don't count, even if we specifically ask for them.

What about money? Surely female animals draw the line at exchanging money for sex, right? After all, money has no value to animals. What would they buy? Well, if they are captive capuchins and they are taught that money can buy some really special items, money suddenly has value. Yes, even to a capuchin. What do capuchins fancy that makes them such quick studies on the almighty dollar? Special food items like grapes, apples, and Jell-O.

Once they had the concept of money down, the capuchins were given a budget and allowed to purchase the items of interest. At some point, maybe because she ran out of money, a female had sex with a male. He paid her promptly and she immediately purchased a grape with her newly acquired token.[7] Whatever you may think about prostitution, however directly or indirectly it occurs, bartering and payment in exchange for goods and services is pretty ubiquitous in animals.

COPYCAT! COPYCAT!

Up to this point, I have discussed some of the qualities females use to directly judge whether a male is good enough for them. Call them lazy, but some females don't worry about independently evaluating males. They take a shortcut, simply picking males that other females have already mated with, essentially copying them. The logic here, presumably, is that some female found this male attractive enough to mate with, therefore he is good enough for me, too. In humans, whether a male is taken is usually easy to determine by the presence or absence of a wedding ring, although some men (and women) remove their wedding band as a mating strategy. Yet with copycats the ring is an aphrodisiac.

Since wedding rings are not observed in other species, how can a girl figure out if her peers have already deemed a male acceptable? If you are a female fish that lays eggs, you can thank external fertilization for solving this conundrum. In many fish, the male guards a little area and hopes that a female finds him sexy enough to lay her eggs there, whereby he promptly fertilizes them and watches out for them until they hatch. Once this has occurred, it becomes clear that at least one female believed he met the criteria of a high-quality male (size, color, territory, sexy spines, and so on). At this point, because he's guarding a nest of eggs, he is basically holding a sign that says, "Someone liked me enough to give me her eggs, so why don't you?"

Indeed, that is exactly what happens. Females will preferentially lay their precious cargo down on a nest that already has eggs.[8] However, there is some limit to how many eggs can be present in a single nest before a female simply will move on and find another male. This has less to do with the worth of the male and more to do with the probability of survival of the offspring if too many eggs are layered on top of each other.

Despite not having external fertilization, mate-copying behavior is also common in animals that have a lek mating system. What is a *lek*? If you remember the nightclub analogy from our discussion about prairie chickens, ladies' night is what you could call a lek system. Males gather in areas where females are likely to show up. Whether the males dance, prance, or honk like hammerhead bats, some females could care less, they just watch and see which male a female chooses and then wait in line.

Surely we are above such a strategy, right? Wrong. It is not merely the stuff of folklore and myth that men in relationships get hit on more often. It is true. Although there are cultural differences, very generally speaking, married men are rated as more attractive, but only for long-term unions.[9] Why is this? And why, if it's a short-term fling she is seeking, is a woman less likely to rate a married man as attractive? If a female only wants a one-night stand, there is nothing to gain from mate copying. His quality as a provider or parent is meaningless in this context. But guys, if you are married with kids, you're apparently the perfect copycat catch for a woman looking for a long-term partner!

Many women out there have experienced this tragedy. It is worse when the situation involves sisters, best friends, or coworkers. For example, my colleague Lisa had a great boyfriend, and she talked to her best friend Abigail about him constantly. Abigail, in turn, would always say how lucky Lisa was to have such an awesome boyfriend. Apparently she thought so much of him that she decided she needed to have him for herself. Abigail argued that it "just happened," but what "just happened" was a clear case of mate copying. The end result was that Lisa lost her boyfriend *and* her best friend.

From a biological standpoint it may not be a bad idea (and, as I've said before, we can't pass biological judgment), but as we are all very well aware, copycat sexual behavior can have a severe ripple effect socially. Despite the consequences, and despite all our reasoning power, it still happens . . . all the time.

Although Lisa learned her lesson the hard way, here's a helpful biological tip for all you young whippersnappers out there: it's a good strategy to wait a while before you introduce your "great" boyfriend to your friends. And it's certainly not a good idea to go around raving about them. If you are in your early twenties, perhaps you should refrain from introducing your boyfriend to your friends at all. Once you hit the mid-thirties and beyond, go for it. Why? Because older females are much less apt to fall into the potential folly of mate copying.

Interestingly, there is still no agreed-upon reason for exactly why this is true, but I have a hunch it may be because older women either (a) are more adept at assessing hard-to-detect qualities like parental-care ability, or (b) know that all relationships require lots of compromise and hard work. These older women realize that while they may win the man, they may not win a "prize."

Ladies, there you have it, it is not just your imagination—other women really are eyeing your husband. So follow the sage advice of Billie Holiday, who once counseled women to be wise, keep their mouths shut, and make sure not to advertise their men.

Gentlemen, this should take some of the mystery out of why women have an instant dislike for young, attractive females being the nanny or your secretary. We are not simply crazy and imagining things—whether we knew it before or just learned it now, we know that these young girls see you as an attractive potential mate, and we know that they are going to be much more likely to act like a cat around you—a copycat, that is! So all you male bosses out there with love-struck secretaries, don't let all the attention go to your head. Keep in mind that her crush on you may be motivated only by the fact that someone *else* thought it was a good idea to marry you.

FEMALE COMPETITION: FROM COPYCATS TO CAT FIGHT

Copying is certainly a timesaving strategy for seeking a mate, especially for young, inexperienced females. But there is another, more nefarious strategy that females have in their toolbox; that is, bootlegging another woman's man. Female competition. And yes, it can get downright ugly. That's right, we are talking about the "cattiness" that can characterize female interactions.

To put female competition into perspective, though, we have to think about what is driving this behavior. Remember, males follow females and females follow resources. Generally speaking, males are competing with other males for access to females, and females are competing for access to resources. Who are females competing against? Other females!

In the most extreme cases, females will outright kill the babies of other females. In some cases, particularly primates, a baby is kidnapped and "aunted" to death. This is where one female, who does not have an offspring, kidnaps another female's baby and keeps it. In the end, because the kidnapper is not lactating, the baby starves to death.

Although not benign, this is not necessarily competition. It depends on who does the kidnapping. When a dominant female takes a subordinate's baby and forcibly prevents the mother from getting her offspring back, you can be assured that it is female rivalry. This is often the case in bonnet macaque monkeys, which I suspect get their name from the bonnet their hair forms on the top of their head.

Anyway, dominant bonnet macaque females seem to have a real problem with subordinate females having babies, and kidnapping rates reflect this. The rate at which dominant females take babies away from subordinates is seven times higher than the rate at which subordinates kidnap the babies of dominant females. In addition to kidnapping, infants of low-ranking females are also attacked, often suffering serious or fatal injuries.[10] Infant killing by females is certainly one way to compete for resources. The idea is that by getting get rid of the hungry mouths to feed, there will be more left for the dominant female's offspring.

Sometimes females lose their offspring as well as their mate. Watching tree swallows in your yard, you might not realize the fierce competition that is going on among females. Typically, a pair of tree swallows finds a suitable nest site and settles down. Both the male and the female participate in raising the offspring and defending the nest. Good nest sites are hard to come by, and there is a tremendous amount of competition for this limited resource. We've all seen the consequences for shoppers on Black Friday when demand exceeds supply. It can get dangerous out there.

The kicker for the tree swallows is that although nest sites are in high demand, males with nest sites are in even greater demand. A female with a nest site is no good without a male to share it with. Sometimes the male conspicuously fails to join his mate in defending the nest. The end result? A female is usurped and evicted from her nest.[11] This is not mate copying. This is outright fighting, with the resident female lying on the ground, too weak to fly, feathers missing, just looking like a hot mess. Aggressive females will also evict a pair from their nest, or even worse, take over a nest that has chicks in it, leaving the newly hatched nestlings to die. For the record, I've personally never seen a shopping fight get that bad.

Not all forms of female competition are as dramatic, but they are just as effective. When several women in a small area, say an office, get pregnant all at once, a joke is usually made about something being in the water. It would be more accurate to say there is something in the air. As a woman, or a man surrounded by women, can attest, get a group of women together and they will start cycling at the same time. This isn't because we consciously try to sync up our cycles; hormones just have that effect on us.

When it comes to competition, some females have a hormonal edge. Pheromones, those sexy sweet smells that attract us to a potential mate, in some species are used against females by other females. Dominant females exert hormonal control over subordinates by using pheromones to suppress reproduction in lower ranking females.

Other than obtaining a complete monopoly over reproductive rights, the benefits for high-ranking females include reduced competition for scarce resources and even free childcare in some cases. In badgers it is all about the food. Unlike North American badgers, European badgers live in groups. I've come up close to a badger, and this is not an animal you want to mess with. Though they are reported to be primarily nocturnal, don't be fooled. They get about during the day, too, and the last thing you want to do is stumble unexpectedly onto a badger.

At any rate, dominant badger females that live in groups have little tolerance for mating by other females when food is in short supply. Not only do they use subtle chemical control over other females, but in instances when their hormones don't quite do the job, they will actively disrupt mating and/or kill the offspring of other females.[12] Badger females have all the bases covered.

Sometimes females compete directly for males, not resources. Where accumulating males is the prize, gender roles are frequently reversed—females will have a harem of males. Even though, like most females of other species, these females mate with multiple males, it is a "special" circumstance. In these cases, the female is flashy, the female displays, and the female runs after the males.

I love the wattled jacana for this very reason, as well as for its name. How can you not simply want to know more about something called a wattled jacana? This is a species of shore bird that makes its home across Panama, Trinidad, and much of South America. Wattled jacana's look like they are walking, or waddling, on water because they walk on the floating vegetation in the wetlands they call home.

Females are larger than males and much more conspicuous, with large, red, fleshy facial ornaments and wing spurs. Such a gender reversal in showiness is often, but not always, a clue that sex roles are reversed and females are in strong competition directly with other females for males rather than for resources. Both males and females defend territories, but females fight with each other to have territories that are larger and surround the smaller territories of several males.[13]

Now why would they want so many males? For these girls, getting more males means having way more offspring because males do all the work. Talk about role reversal! Males incubate the eggs and do most of the child rearing. Wattled jacana males embody the Mister Mom persona, freeing females up to mate with additional males. Not that the females have it easy. They spend a lot of their time and energy trying to convince new males to mate with them. On top of that, they must constantly be on guard, ready to fight off females.

Human female competition can also get physical, and it is no less vicious. By now, we've established that women are, for the most part, competing and getting aggressive over two things: men and resources. One argument for why human female-female physical combat is less prevalent than male-male combat is that fighting is risky. There is always a chance of getting hurt, and when women assess the cost and benefits, they realize that injury could impair their ability to care for their offspring, which would in turn lower their children's odds of survival.[14]

Another theory is that it is less socially acceptable for women to get into a fistfight. So what do we women do instead? Our weapons are usually not our fists but our mouths (I can feel all the men nodding their heads right now). Around the globe women yell, scream, shriek, insult, and belittle their opponent. Relations between the sexes are pretty aggressive in Kashmiri society, and the women are champions of verbal abuse.[15] For the women of Ona, verbal duels can last for hours.[16]

It would seem we women have a long history of perfecting the art of giving out a tongue-lashing, and this propensity may explain why few, if any, men can actually win a verbal argument with a woman. As an aside, recent studies silence the long-held myth that women talk more than men. The results clearly show that while men and women may talk about different things, both sexes utter roughly the same number of words.[17]

But I digress. Bickering, gossiping, and "informational warfare" are additional tools of choice. And what better tactic than to go after another woman's reputation?

That does not, however, mean that women are above physical assault. Whether we bite, scratch, throw things (a female favorite), or destroy the property of a rival, we are willing to go after the female we perceive as a threat. And back to the cost-benefit balance, this is far less risky than going after a man. Although, as we will see, controversial studies have shown that the rate of female physical assault against men they are dating is much higher than the rate at which men attack women they are dating.[18]

MOVIN' ON UP

Getting and keeping access to better quality mates and resources is desirable for females of all species. Yet it is not always efficient to engage in so much competition. To save a little energy, there is an alternative tactic available. As the copycat saves time in deciding who's a good mate, this approach also saves energy in the "searching for the best mate" department.

Have you ever heard the phrase, "If you settle for less than what you deserve, you end up with less than what you settled for?" For some female animals, the phrase would read, "Settle for less than the best right now, because you can always take better when it comes along."

Suppose you are an adzuki bean beetle female. You would measure about three to four millimeters in length and look, well, like an adzuki bean. Adzuki beans are common in Asian cuisine in the form of red bean paste, which thankfully does not include the adzuki bean beetle. Maybe. Since females lay their eggs on the bean and new beetles "hatch" from inside the bean, you never know.

If you were born with a lot of siblings in your bean, you may end up being a small female. So here you are, a small female, ready to mate, and as you go along, you encounter several males. Each time you encounter a male, you ask yourself this question, "Should I mate with him or pass?" After all, if you pass, you may not encounter another available male. Sort of like living in Jacksonville, Florida, or El Paso, Texas, where single males are in short supply. What's a girl to do? These small girls will lower their standards and mate with a small male, but should a better one come along—oops, my mistake—they have no problem trading up to play with the big boys.[19] Essentially, if a female doesn't encounter multiple males simultaneously and instead comes across them sequentially, one option is to mate with the first male she encounters (to ensure reproduction), and then get pickier and pickier about the next male, and the next, and so on. . . .

A similar scenario plays out in guppies. Picture this: a virgin guppy female is given the choice of two males who vary in attractiveness. I know, you are wondering what exactly turns a female guppy on. You've got to have large patches of orange. Seriously. Okay, back to the experiment. Virgin female; two males. Let's say male #1 is so-so and #2 is hot, hot, hot. If said virgin is presented with male #1 first, she mates with him. Then, when presented with male #2, she mates with him also. However, if you reverse the order and present her with the "oh, caliente" male #2 she automatically mates with him but then subsequently rejects male #1. Guppy females never trade *down*, only up.[20]

By now the wheels are turning and you are starting to wonder if human females engage in similar behavior. It's not just a myth, we do. And we do it like guppies, except we don't care for orange patches on the skin. How do we determine who is "*oh caliente*"? Congratulations to those of you who said we use sex. Yes, human females use sexual infidelity as a tool to "sample" additional males and determine if they are worth trading for.[21]

Sometimes women, like my friend Susan, only find out after they have gotten divorced just what they were missing out on for all those years. Susan had married her high-school sweetheart, but once she was back on the dating scene, let's just say she discovered a lot by using a very scientific sampling approach. Notably, what she learned is that she had been married to an inferior male. Now, sexual performance ranks high on her list of qualities as she chooses a new mate.

GIRLS JUST WANNA HAVE . . . A LITTLE VARIETY, PLEASE

This brings us to a critical juncture. We will talk specifically about monogamy, or lack thereof, further on, but here is a teaser. To believe that boys will be boys and girls are looking for the "one" is like saying you believe in unicorns. Though this idea has been imparted for ages, the truth is that females, more so sometimes than males, biologically benefit from having multiple partners.

Indeed, after DNA and genetic testing became available, a lot of head scratching among scientists (mostly male) ensued as they tried to figure out what the heck was going on. The reason? It looked like female promiscuity was rampant. If males mate with multiple females, why was it so surprising to find out that females mated with multiple males? With whom did scientists think males were mating? Unicorns?

One of my favorite species of frog is the strawberry poison dart frog. Members of this species can come in many varieties, but possibly the most recognizable is the blue-jeans kind, which has a red body with blue legs. They are so tiny you might miss them. Males are vigorously territorial, but females could care less about the males' territory, body size, or coloring. Nope. This is a species of convenience. A male only needs to meet one crucial requirement: be nearby.

Now if you are a male who has no other competitor within a five-meter radius of the female, then she is all yours. Otherwise, a female will mate with all the males in close proximity.[22] This begs the question, if females are not trading up, why are they mating with multiple males? One of the best, indirect benefits a female can get is simply hedging her bets. Her genetic bets, that is. Diversifying your investments, if you will.

We all know the phrase, "Don't put all your eggs in one basket," and this cannot be truer than for your actual eggs.

A female takes a much greater risk letting just one male fertilize all her eggs. What if he has crummy genes? What if he's a lousy lover? The solution is simple. Reduce your risk by having several males fertilize your eggs.

I love snakes like I love oceans—from a distance. Indeed, after expounding on all my mishaps with the ocean, a very famous coral reef researcher told me I should call him before undertaking any activity involving the ocean to assess its safety.

Fortunately, I have had no such traumatic encounters with snakes. Tropical water pythons belie their name, as they are known to wander away from water. Found all over tropical Australia, they can grow up to nine feet long. Females that mate with multiple males have offspring with greater genetic diversity and greater survival than those that only mate with one male.[23] This is true for too many species to list. Along the same lines, some females don't get a genetic benefit; instead, they simply produce more offspring as they mate with more males. The irony here is that this is the very same reason given for why males mate with many females: to produce more offspring. Hmmm.

Sometimes mating with multiple males is a way to stock up on sperm. After all, it may pay to hold on to sperm for a rainy day, like when you are actually ovulating. Females of most reptiles, likely all birds, and yes, even mammals, can collect and store sperm from many males and then dole it out (internally) as they please. Another neat trick is to delay implantation. Depending on the species, the sperm can be kept viable for a few hours or several years. This explains some but not all cases of apparent "virgin" births, or parthenogenesis, particularly in reptiles.

The Javan wart snake, a large water snake found in India, holds the current record at seven and a half years.[24] Do human females store sperm? Not exactly, but like females in many other species, the exact timing of our ovulation is obscure, leaving males no way to pinpoint precisely when we are fertile. Once the egg is released, there is a very narrow window of time that during which fertilization can take place. By concealing ovulation we can obfuscate paternity.

Even more important, because human sperm can live anywhere from three to seven days, depending on the environment, all males that a woman has sex with during this time can potentially be the father.[25] This strongly implies that there is a benefit to multiple mating by human females. What is the benefit? One possibility is that it's best to settle down with a male that has good resources and then mate with additional males that have superior genes.

While we don't directly store sperm, how much sperm a woman retains is tied to whether or not she is having an affair. Yes, fellows, you may have to read that sentence again. As if that wasn't enough, let's drive the point home. Women who are having affairs unconsciously time those affairs to coincide with when they are ovulating. And just to make it absolutely clear, women who are having affairs have more orgasms with their lover than with their husband.[26]

Given the costs of paternity in species where males provide paternal care, males have developed counterstrategies to circumvent a female's desire to confuse paternity. But why would it be advantageous for a female to conceal true paternity? Beyond simply securing additional resources, a good father, or better genes, there are some instances when it is necessary for the female to confuse all the males around her about who exactly the father is—for instance, in species where males have a tendency to kill babies that are not theirs.

Therefore, one common reason females mate with multiple males is to protect their offspring from infanticide. If you think about it, it makes perfect sense. If a male mates with a female but cannot know whether he is the father, he is less likely to be aggressive to that offspring and even less likely to kill it.

Even though this happens in a wide range of primates and other mammals, let's talk more specifically about langurs. When you look at a picture of a langur, its facial expression gives the impression that it is pretty ticked off. If you are a female langur this is completely understandable, since you have males running around killing your babies.

Common to India, the gray langur, or Hanuman langur is among the best-studied langurs. A female langur lives in a group with one or more males, and she also has a tendency to mate with more than one male, and usually with males that don't belong to her group. This is because males from outside the group are almost always the ones who come in and try to kill langur babies. The DNA evidence confirms that confusing males about whether they could be the father works. The male langur simply does not kill his own offspring, and if he mated with a particular female, there is a chance that her baby could be his.[27]

Having many confused males around you could be a good thing if infanticide is going on, but there may be other benefits to having multiple males around. Each male can serve a different purpose or provide additional resources that a female needs. Remember Susan, the newly minted divorcée who began sampling the goods? She has concluded that she needs five men. One who can satisfy her needs in bed; one who is handy around the house; one who loves to go shopping; one who is romantic; and finally, one who can be her best friend.

Whew! That is an exhaustive list, but she claims it is why she continues to date multiple men simultaneously. They all fulfill different needs. She is not alone. As I mentioned earlier, baboon females like having male friends, too. In chacma baboons, these males will groom the female, baby sit for her, and come to her assistance in times of trouble. And guess what? Females compete for these male friends!

BOY TOY, ANYONE? THE ALLURE OF THE YOUNGER MALE

If you read the news or watch television, you might have the impression that younger females have been recently thrown a curve ball when competing for partners. Now they have to compete not only against young women like themselves, but against older women as well. Hello, Mrs. Robinson. Today, a woman may be called a cougar in jest, or more frequently, as an insult by younger females who are mystified as to why young, sexy males would be interested in older (more experienced) females. Who can blame them? Their whole lives, these young women have been told that youth is everything; it is the elixir of life.

But this is not a new phenomenon at all. The French have known all along about the benefits of older women, and French culture is one in which women can be seen as sexier as they age. It makes me want to be French. The word "cougar" implies that older women go hunting for younger men, but the reality is that, in many cases, younger men are pursuing older women. In the next chapter, we will take a closer look at why these young males prefer older females, but for now let's focus our attention on what is in it for ladies.

There is a common perception that the quality of a male goes up with age while the quality of the aging female declines. Rapidly. It is certainly true that in a suite of species females do prefer older males. Older males are stronger, bigger, more experienced—all of which enhances their value in the eyes of females.

Heck, just by *making* it to old age, a male can demonstrate he is of superior quality. It is a dangerous world out there, more so if you are male. All that aggression, not to mention the energy it takes to be flashy, constantly displaying, and proving your sexual prowess. It's exhausting and comes at a cost that sometimes makes longevity a signal of quality. There is some evidence for this in birds and a few other species. In a nutshell, older males often have the goods and can deliver them.

Or do they? The biggest benefit to females from mating with younger males may just be the quality of their genes. In the wild, as we already know, keeping in mind our criteria list, genes trump virtually everything. Work on fruit flies and sand-flies tells us that the offspring of older males do not fare as well as those fathered by younger males. What's going on here? The biggest problem with older males may be their sperm. There is a lot of talk about how the quality of a female's eggs decline with age, but, until very recently, no one was talking much about aging sperm. Why is that, by the way?

The three main drawbacks to mating with an older male all have to do with sperm quality. As time goes on, genetic mutations accumulate, fertility declines, and subsequently the reproductive quality of ejaculate goes down. The latter is important if females gain some nutritional benefit from the ejaculate. Wait, there is a fourth—older males are less adaptable to changing conditions. The firmly set-in-his-ways old bachelor who likes things just so comes to mind.

Setting aside their lack of adaptability, older males make better, more experienced fathers, right? Not necessarily. In humans, the evidence is accumulating that the sperm of older males is of poorer quality, and the children born to older males may have a shorter life span.[28]

Clearly, male age is important when it comes to genetic quality, and we already know that females use a variety of traits to assess male genetic quality, including stamina. Remember the dance of the prairie chicken male and the display that goes on for two months? No one has looked at age preferences in prairie chickens, but researchers have looked at age preference in crickets. Female crickets prefer the younger boys because they have better stamina for, um, singing. We have little to go on in terms of research studies, but based on Susan's experience, I suspect this may play a role in partnerships between older females and younger males in humans, since not a lot of offspring are produced from such pairings.

As I mentioned earlier, Susan was thrust back into the single world in her late thirties. At one point, she relayed to me her surprise at the number of younger men hitting on her, ranging in age from eighteen to twenty-six. It was ridiculously flattering and boosted her confidence to a new high. They all seemed eager to convince her that they could more than take care of her needs.

When I asked her if she partook of the fountain of youth, she said she rebuffed most of them. She knew exactly what she was looking for (like many older women do), and according to her, the much younger men hitting on her just wouldn't be able to meet her more "seasoned" list of criteria—both in and out of bed. She was also well aware that young men have "heard" that women in her age group are sex-crazed and, not surprisingly, they find this very appealing. There is truth to this, as sexual desire peaks for most men in their teens and for most women later, in their thirties and forties.[29] An older woman who chooses a younger man is often selecting him mostly based on aesthetic criteria and much less for his territory and resources. Whether consciously or not, these older women (for the most part) are not planning on bearing offspring with the younger men, so their criteria in choosing them changes.

Susan did not rebuff them all, however. There was James. James was twenty-three and, according to her, beautiful. Body "like a Greek God." She said that right away there was something unusual about him. He had confidence, but not in an arrogant way. She wasn't sure how they ended up talking, but he didn't have the same eager look in his eye as the other ones. He was different, and she was attracted to different. She told me about some of the things she liked about James: how much undivided attention he gave her, the unhurried way he talked to her, that he would take her for ice cream and out to dinner, that he would cook for her and hold her hand. She said she really felt valued and appreciated. And then there was the bedroom.

As you men probably already know, we women talk about it, or, I should say, we talk about you and your performance. All I will say is that after she relayed their first encounter, I needed a fan and a big glass of water. It was the stuff of romance novels. Not only was James sweet and gorgeous, but she effectively had to throw out all her preconceived notions about younger men not knowing what they are doing, and he wanted nothing more than to make her happy.

Surely they got married and lived happily ever after, right? Not so fast. The long-term survival rate for such relationships is fairly low. My suspicion is that the biggest challenge is not the age difference per se, but the lower reproductive potential of older women. In the case of Susan and James, she really wanted to have a baby right away, and they both knew that babies were not going to be part of James's plans for years.

TICK-TOCK: IS THERE REALLY A CLOCK?

This brings us to the inevitable question of the biological clock. I will never forget a date who asked me my age and then, when I responded, "Thirty-eight," he replied, "Wow! Tick-tock, tick-tock, right?" As you might be able to already guess, there was not a second date, but I did leave him with this thought. Sperm is relatively cheap and not that hard to come by. Heck, we can even get it from a bank.

Okay, I was a more than a little put off, but the truth is, we do not necessarily need a man present to have a child, whereas men definitely need a woman to accomplish the same task. That being said, there is an inescapable time crunch, and it doesn't do us any good to deny it. Despite more and more women having children later in life, even as late as their fifties, it is not without considerable difficulty and risk.

The fertility rates of forty-five-year-old women are only about 10 percent while the miscarriage rate is almost 90 percent. Even if conception occurs and pregnancy succeeds, the risk of genetic abnormalities in this age class goes up to one in twenty-one.[30] Coupled with risks to the mom, higher rates of stillbirths and premature babies mean that there is no good biological argument for delaying reproduction into the mid-forties and beyond. The exception seems to be women who harvest their eggs in their twenties and thirties and use in vitro fertilization to achieve pregnancy.

One of the most well-known chimpanzee moms of Jane Goodall's long-term study was Flo. Flo was approximately forty-five when she died. I religiously watched all the National Geographic videos documenting the dramas of the chimpanzees of Gombe, dreaming one day that I might get to visit them. I distinctly remember that, in her prime, Flo's children flourished. Faban and Figan both went on to be alpha males, while her daughter Fifi died at the ripe old age of forty-five in 2004.

It was the children Flo had in later years that did not do so well. She was in her later thirties when she had Flint, and she was likely just forty when she had Flame. Flame died at six months old, most likely because the older Flint was never properly weaned and "stole" valuable resources from his younger sibling. From the documentary, it seemed that poor Flo was just too tired to discipline Flint. Although he should have been largely independent by the age of eight, Flint died only one month after his mom.

My guess is that, had she lived longer, Flo would have surely hit menopause, if she hadn't already. There are few animals out there that live long enough to reach menopause, but we join ranks with chimpanzees, macaques, elephants, and pilot and killer whales in living well beyond our last pregnancy.[31]

Regardless of whether the females of other species hit physiological menopause, as we know it, it is clear that for us there is a clock, and it starts ticking when we are in our mid-twenties. This biological clock, combined with the emotional desire to have children, can send some human females into a dangerous desperate spiral. The irony is that in the wild, because males usually court the female, being a desperate female that chases after males is not only unnecessary, but may actually be a signal to a male that you aren't a very good mate.

WHEN THINGS GO TERRIBLY WRONG: THE MALADAPTED FEMALE

The desperate female may engage in many maladaptive behaviors, or behaviors that don't make sense. Movies such as *Fatal Attraction* have been made about this, striking fear into the hearts of men everywhere. There is deceiving, stalking, maiming, and killing, to name just a few maladaptive behaviors. This desperation may be to acquire a mate, keep a mate, and sometimes even to punish a mate for straying or leaving.

When it comes to stalking, both men and women do it, though the motivation for the behavior differs between the sexes. More often than not, women stalk someone they already know with whom they have a desire either to become sexually intimate or to remain sexually intimate. Twenty-five percent of the time female stalking behavior is because women don't want the relationship to be over. When a woman stalks a man she is more likely to use the phone rather than to physically follow him—what one of my friends refers to as stalker dialing. But female stalkers are also just as likely to inflict physical damage to property or person. The image comes to mind of an enraged girlfriend destroying her boyfriend's car or slashing his tires because she caught him cheating.

When it comes to violence, contrary to popular belief, studies show that human females are often more likely than men to physically assault a dating partner, and that assault is also more likely to be severe.[32] However, it shouldn't be a surprise that the rate of injury inflicted by females on males is much lower. This may simply reflect differences in physical strength between men and women, though some women go so far as to maim or kill their mate in retaliation for ending a relationship or, more frequently, for cheating.

Do we see this kind of female aggression and violence in animals? Though most female-to-male aggression in the wild is not motivated by jealousy or retaliation for ending a relationship, female animals can *absolutely* be physically aggressive toward males. Female Japanese macaque monkeys use aggression to limit the number of males that can be part of the group, and Milne-Edward sifaka lemur females will be extremely aggressive toward outside males or when they want to oust an older male from the group.[33]

That doesn't mean there isn't sexual jealousy. Like the tree swallows mentioned earlier, female red-winged blackbirds stand to lose a great deal if another female filches their mate. As a result, they will even attack a fake female that is placed in a sexually solicitous position.[34] Clearly, female animals will most definitely use aggression to safeguard their mate from another female, but here we see that they usually attack the female, not the male. This is remarkably different from what we see in humans, where women attack not only other women but also frequently men. Indeed, if you account for the lower rate of violent crimes women commit in general, you find that the rate of mate killing is not notably different between men and women. What is going on here? Do other females kill their mates?

With few exceptions, including that of cross-species deception like the firefly females who lure males of another species to their death, female animals generally do not lure, stalk, harass, maim, or kill potential or actual mates. The most notable exception is spiders, including the black widow spider, the red-backed spider, and the orb-weaver spider. Life is pretty rough if you are a male spider trying to mate with a female. There are a lot of things conspiring against you. Even if you manage to get it all right, you risk serious injury that may prevent you from mating ever again. Oh, and she still might eat you.

As you may already know, the most well-known of this group of at-risk spiders is the black widow. Movies have been made dramatizing the real-world mating consequences for males: being killed and eaten by your mate. For some time it was believed that it was advantageous for the male to get eaten, with some researchers even going so far as to suggest that males willingly commit suicide.

However, recent research on the red-backed spider, a relative of the black widow, reveals that males that get eaten do not father more offspring, thus they are not getting any real reproductive advantage for their suicidal tendencies.[35] The leading current explanation for this clearly maladaptive behavior is that members of this group of spiders are asocial, extremely aggressive, and just generally don't like each other much. As a result, there is a spillover effect causing females to eat males, some of them even attacking every male they encounter. Therefore, simply coming together to mate is a hostile affair, and males are on the lookout for aggressive behavior by females when they are attempting to mate. While these poor male spiders can't avoid aggressive females, since they are all pretty pissed off, human males certainly can stay away from angry women. Fatal attraction should be left to the spiders and praying mantises, and men would be wise to proceed with caution should they detect any of the telltale signs of a maladaptive woman.

While men are definitely better off steering clear of angry mates, especially as long-term partners, it is women who have evolved to be the choosier sex, and according to evolutionary biology, they *should* be extremely choosy when selecting a mate. There are various traits that indicate male quality, and females, whether they are aware of it or not, pay very attention close to these. Whether it is looks, stamina, resources, genes, or some combination thereof, these traits tell females something very important about a potential mate. Can you imagine a world in which human males dance for us, dress up for us, show off their resources, and bring us presents, all in an effort to convince us to mate with them? Oh, wait, we do live in that world, and it is actually all that female choosiness that has created it.

6

PEACOCKS, LIONS, AND MEN

I have come to believe that women might just be giving men the short end of the stick. Could there possibly be a general lack of appreciation for men? As a woman, I have decided it was fair to reflect on this. All this came after a conversation I had with one of my best male friends, Oscar. We dated when I was sixteen; he was my first boyfriend, and we have remained friends ever since. After not seeing each other for ten years he came to visit me. As the days went by I started noticing things about him that surprised me. There was also a different texture to our talks.

One evening, at the kitchen table, I had a sudden revelation, "I have known you for over twenty years and yet it seems like I know nothing about you," I said, "Why is that?" His answer humbled me and forever shifted my perspective on males. He said, "You never asked and you didn't seem interested. It was like you decided I was supposed to be this way or that way simply because I was a guy, and there was no room left to just be myself." He said it without judgment or criticism, as only a true friend can do.

Suddenly, it dawned on me that maybe it was difficult to be male. I mean, don't get me wrong, it's no walk in the park being female, but this was truly an "aha" moment. It had never occurred to me that it might be hard being male. It can be easier to step out of my role as a biologist than my role as a female, but I really wanted to get out of my comfort zone and look at what it means to be male, to examine the cost of being male, and to try to understand the biology behind male behavior. So, as a biologist *and* a woman, I began to wonder how things look from their perspective.

THE COST OF BEING MALE

During my professional training I had certainly learned that males on average die younger than females. This is true for most species, including humans. But suddenly this information was of great interest to me. Now I wanted to understand why. One idea is "live fast, die young," which suggests males trade reproductive effort and success for a shorter life span.

In general, since males start breeding later and stop breeding sooner than females, they have a narrower window of opportunity to reproduce. Not quite the forty-year-old-virgin syndrome, but in some cases it can get extreme. In Australia there are several species of marsupials in which males age so fast they don't make it past a single breeding season.

The carnivorous marsupial, the northern quoll, for example, embodies life in the fast lane. In the winter, when females are ready to mate, there is no elaborate courtship, no romantic dinners. Males just grab a female and, if she agrees, mate for upward of twenty-four hours before finding another female and starting all over. Sounds like a dream right? Not really for the northern quoll. The energetic costs of mating for hours and days on end are severe. Weight loss, anemia, hair loss, and parasite infestation—which then usually leads to "complete post-mating mortality" within two weeks.[1] Translation: *All the males die.*

Granted this is quite extreme and also relatively rare. What is not unusual is that males often compete physically with other males. Fighting with other males is serious business, and the "high risk, high gain" hypothesis suggests that adolescent and adult males will be more likely to die as a consequence of such "risky" behavior.

I will never forget visiting Kruger National Park in northeastern South Africa. I saw many amazing animals and had some unbelievable experiences, which included a mischievous vervet monkey stealing my sandwich. One species that I saw was the kudu. With white stripes that look like paint running down the sides, a stripe connecting the eyes, a milk mustache, and a white beard, the kudu is a beautiful antelope found over eastern and southern Africa.

In any given population, there is a one-to-one sex ratio of males to females at birth. The males have gorgeous curved horns, which they lock with one another when they are fighting. By the time they are six years old, males have only a 50 percent chance of making it to the next year, while female survival goes up year after year. What gives?

Male kudus have a lot of things working against them. They have to grow bigger, and this takes more energy. Just think of a gang of hungry teenage boys raiding your refrigerator. They can clean you out. Upon reaching adulthood, men have, on average, 60 percent more muscle mass than women. It takes a lot more food to grow all that muscle. Male kudus also have to eat more to meet the energy demands of being a dude. When food is in short supply, they are more likely to starve.

Another real danger for these males is dispersing from home. This is a common hazard for males of many species. Sometimes they form a gang with other single males, but usually they go it alone. When you go it alone, it puts you at greater risk for being eaten. Lions seem to prefer larger kudu, getting more bang for their buck, so to speak. As a result, lions target male kudus more often than the smaller females.[2]

Even though young human males aren't too likely to be eaten by a hungry lion, it is still far more dangerous to be a man than a woman. Young adult males are two to three times more likely to die. The leading causes of mortality in young men in the United States are motor vehicle accidents and homicides.

After car wrecks and homicides, suicide is the third leading cause of death, with rates again higher for males. Almost across the board, whether it is tobacco use, violence, or drug and alcohol use, young males take the prize for engaging in riskier behavior.[3]

Presumably once a male makes it past puberty and young adulthood things should stabilize, right? Not exactly. Even after controlling for risk-taking behavior in adolescence and young adulthood, males are more vulnerable to infection, injury, stress, physical challenge, and degenerative diseases, all of which contribute to earlier death.

Why does being male make you more vulnerable? Right out of the gate males may just be more "fragile" than females, from a developmental and physiological perspective, that is. The elevated levels of testosterone, that lovely intoxicating hormone, may actually contribute to lower functioning of the immune system, which may make males more vulnerable to disease. In the United States, excluding HIV infection, mortality as a result of infection is 30 percent higher in men.[4]

One way for men to extend their life is to be married. Whether this is because being married to us ladies makes you happy or because we make sure you get to the doctor is still up in the air. In 2012 there was an exception to this pattern in Italy, where married men were committing suicide at an alarmingly high rate.[5] As I already mentioned, men are more likely than women to commit suicide, and in Italy it is usually unmarried or widowed men who kill themselves.

So what happened in 2012? The Italian government turned tax collection over to a private company, and the aggressive new tax-collection practices, combined with poor economic conditions, led to an uptick in male suicide by married men. Many of these men were consulting attorneys prior to committing suicide to ensure that the wives and children they were leaving behind would not be burdened with their debt. Why were so many married Italian men committing suicide due to financial problems? Perhaps it was because a more conservative and traditional belief in gender roles made the Italian men's inability to provide for their families intolerable. And this phenomenon is not limited to Italy. India's farmer suicides have received international attention, with male suicide rates increasing drastically in recent years for much the same reason as in Italy.

Men have traditionally been the ones who fight, go to war, and bear the brunt of the stress associated with acquiring enough resources to secure and raise a family. As if that weren't enough, we think of men as not being subjected to the deep ocean of emotions reserved just for women.

Although new studies are showing that boys might actually be more sensitive than girls,[6] we still generally believe that men have to be strong and stoic—the pillar on which everything else rests. This is the picture we paint of males, but the reality is that the cost of being male can be steep.

GETTING THE GIRL OF YOUR DREAMS—CAN I BUY YOUR LOVE?

When it comes to getting the girl, males have an arsenal of strategies at their disposal. I have already spent a lot of time on how males fight. And it pays off. One surefire strategy is for males to get big, be flashy, and directly compete in fights or displays with other males for the attention of females. We get it. Yawn. What else have you guys got up your sleeves?

We've seen how females will gladly exchange sex for stones, meat, or money. Like their human counterparts, males of some species specialize in gift giving. In many spiders, butterflies, and crickets, food is the gift of choice as a way to entice the ladies. This is referred to as a nuptial gift. For the nursery web spider, offering females a meal is par for the course. A delicate insect wrapped in silk is delivered to the female. Dinner anyone? Now you might be inclined to think that the silk wrapping is there to make the female happy or to make her think it is fancy, but the truth is that males wrap the present to reduce the chance that females will just take the gift and run off. Kind of like when men buy a woman a drink at the bar and then she mysteriously has to "go to the bathroom," taking her drink with her. By wrapping it in silk, the male ensures that the female has to open it, which takes time, reducing the chances she can grab and go, and in turn giving him more time to mate with her.

If, by chance, the female tries to grab the gift and run, the male plays dead while maintaining a death grip on the wrapped insect. The female may run off, but she drags the male along behind her. Eventually she stops, and he magically comes back to life and pounces on her to mate.[7] Talk about faking it! Perhaps this is why some men think that dinner and a movie are enough to convince a woman to mate. After all, they did feed you. Did the third-date rule for determining whether a couple will proceed to sexual intimacy evolve to protect males from females who just take the meals and run? I wonder. . . .

Some males try to mitigate the cost of a gift by giving a fake or empty gift. Some male nursery web spiders offer females a worthless gift rather than a meal. The fact that they wrap their gifts in silk actually helps them cheat. Most of the time females figure this out before agreeing to mate, but occasionally they are caught off guard. Males will catch an insect, eat it, and then wrap the skeleton in silk and present that to the ladies. Sometimes they even go so far as to present a female with a beautifully wrapped inedible object, like a plant seed. By the way, spiders don't eat seeds—ever.

And it happens about 40 percent of the time. Like the guy who says he doesn't take women on dates, male nursery web spiders that don't give gifts don't mate nearly as often as those that do. This provides an incentive for trying to get away with giving a worthless gift. By the time she figures out that she got an empty gift the male has already begun mating with her.

Does this ever happen in humans? It does, and it reminds me of Gina. She was involved with this guy, who, from the get-go, I did not particularly like. Despite the poor first impression he made, she insisted I give him a second chance. Which I did. Until the chintzy-gift incident.

Before things became sexually intimate between the two, he showered her with gifts. Clothes, fancy meals, and jewelry. Gina's not the type to relish lavish gifts, but he assured her that he loved her and enjoyed spoiling her. Then I got the call. This "boyfriend" had given her a key ring for her birthday. And no, it did not include the keys to a brand-new car. It was just a key ring.

Her first instinct was to think, *Okay, I'm not a materialistic person; it's a special key ring because he chose it for me*. That is, until she saw the inscription on the back. I know . . . your first thought was that it had the name of another woman on it. It was worse than that. It was a door prize given to everyone who had been to the local casino at some point the week before. That's gotta sting.

For Gina, it wasn't so much the gift itself, or the fact that he was out gambling instead of spending time with her. Rather, it was everything that had preceded it. What had happened to how she "deserved to be spoiled"? It only made it obvious that his lavish generosity had been to secure the goods. My suggestion was to do what nursery web spider females do when they detect chintzy gifts: terminate copulation immediately. This could be the upside or downside of having an animal behaviorist as your friend. I guess it depends on your perspective.

Now some guys may consider their sperm a gift, but in field crickets sperm really is. And it is a gift that keeps on giving. When sperm is the gift, it is called a *spermatophore*. This spermatophore includes sperm and seminal fluid, which contains nutrition for the female. This is pretty costly for the male to produce because it can weigh as much as 30 percent of his body weight, with protein making up as much as 10 percent of the content. The male places his gift on the female's genitals, and after a reasonable amount of time (presumably for his swimmers to get moving), she takes it off and eats it.[8] Yum.

If spermatophores aren't your thing, males can also provide more personal services, like attention, grooming, or a massage. I am quite partial to these types of special attention. For that reason, I wouldn't necessarily mind being a monkey, say a long-tailed macaque female, since males engage in long periods of grooming with their chosen lady. The male orb-weaver spider also performs a carefully executed backrub. However, if he fails to massage his girl just right, she eats him. No pressure, right? Fortunately, if a human male takes a woman to a bad restaurant, offers her a regifted door prize, or gives her a crappy massage, she doesn't devour him. A ruthless tongue-lashing may be in order, but not death.

ALPHA, BETA, GAMMA—IT'S NOT EASY BEING ON TOP

They say power is intoxicating, and for many species that means one thing, and one thing only. Achieve alpha-male status. If you are an elephant seal alpha male you will mate with over 90 percent of the available females on your beach. Elephant seal males are massive and disproportionately larger than their female counterparts. Weighing in at between five thousand and eight thousand pounds, with a giant nose, males also, not surprisingly, don't live nearly as long as females. Some subordinate males do manage to get a few females here and there, but by and large it is the dominant male that takes all. The trade-off in elephant seals is that alpha-male status lasts only one or two seasons before the alpha is unceremoniously replaced by stronger, better males the following year,[9] kind of like some high school jocks who have one or two good years and then fizzle out, leaving nothing but a floppy belly and a defeated look in the eye.

Believe it or not, some alphas are actually insecure about their status. You can usually spot these types a mile away. They are always yelling, blustering about, and otherwise making a fuss. It is understandable to some degree. After all, it is stressful being at the top. There are enemies everywhere, since other males are always trying to dethrone the alpha. It could make any top dog a little neurotic.

Maybe this was what was wrong with Frodo, one of the chimpanzees observed by Jane Goodall. Frodo's older brother, Freud, was a confident peaceful alpha, whereas Frodo seemed plagued by the need to continually and aggressively defend his status. Blood was not thicker than water in this case, as Frodo deposed Freud while he was ill. Frodo was the largest male, and you would think that he wouldn't have needed to be so aggressive. Yet his attacks were not directed only toward other chimpanzees; they were also sometimes directed toward people. Though, at some level, these types of alphas must fight to remain on top, they also must maintain a delicate balance. For example, in chimpanzee communities males need political support and alliances to maintain their position. That means that you can't just go around beating everyone up.

This brings up an interesting thought. Why would another male support someone else in being alpha instead of just trying to be an alpha himself? The obvious answer is that, for whatever reason, he is not cut out to be an alpha male and is better off being a wingman. And there are perks to being the wingman. For chimpanzee males, and probably human ones as well, this means sex—though not exactly in the same way. For chimpanzees, in exchange for political support from lower ranking males, the alpha male may tolerate a wingman mating with "his" females.[10]

Unlike Freud and Frodo, some brothers will share mating opportunities because it can be hard to get to the top alone. This happens quite often in lions. But lions also form coalitions, with anywhere from two to ten males that are not brothers. These gangs of bachelor males work together and take over a pride of females, ousting the current alpha male or males.[11]

Why would these lions share alpha status? Ultimately, it's again for sex. A coalition of males is more successful at exiling the existing alpha(s), especially if the coalition only has to evict one male. More importantly, coalitions of males can access multiple groups of females. This translates into more mating opportunities for each male than he might get alone. It is almost like the hunting technique used by lionesses. By cooperatively hunting, lions can take down bigger prey, which equals more per-capita food than if they hunted alone. However, once the coalition has deposed the old alpha, the males within the team then hash it out for mating rights, which means that there still may be an alpha among all those alphas! It's enough to make your head spin.

The term *alpha male* may be a household name, but few know of the beta and gamma males out there. These different types of males reflect alternate mating strategies. Beta males come in many varieties, either varying in physical appearance, behavioral characteristics, or both. Instead of fighting for the top spot through traditional methods, they have a way of "sneaking" in through the side door.

Alternative mating tactics are found in many organisms, from crustaceans to mammals. One popular strategy among crustaceans, fish, and octopuses is to be a sneaker male. And yes, it is *octopuses* not *octopi*; Greek derivation, not Latin. These males are smaller and less aggressive than their alpha counterparts, and they may go unnoticed by the larger dominant males. Sometimes they even imitate the behavior of females.

I remember as a kid being fascinated with octopuses. They

are expert chameleons, deftly using camouflage to hide from their predators. If they are seen, they either zip on out of there at lightning speed or combine a speedy escape with a well-placed ink squirt. They are incredibly smart, too; they are considered the most intelligent invertebrate. They have impressive cognitive capacities, short- and long-term memory, and a highly complex nervous system, which makes it barbaric that some countries still allow surgical procedures on octopuses without anesthesia.

With superior problem-solving skills, including unscrewing jars, it is not surprising that some Indonesian male octopuses have figured a way around all the fighting and stress that goes along with trying to fend off other male competitors. Rather than waste their energy on all that, some males become sneaky beta males. A female and male couple will generally share adjoining rooms, or dens. The male is pretty serious about making sure no other males get access to his girl. Males have stripes that signal their manhood to others. Smaller, underhanded beta males will masquerade as females, going so far as to conceal their stripes while swimming past the larger male. You learned it here first—octopus transvestites.

The deviousness doesn't stop there. Rather than blatantly canoodling with the female, the intruder may hide behind something like a rock and extend his mating arm in search of the female. Sometimes this strategy can backfire, as the alpha male may begin to fancy this new "female" and try to mate with him![12]

I know, I know, you are probably thinking, *Wait! Didn't she tell me that animals don't lie about their appearance to secure a mate?* Okay, you got me. Sort of. The reason male octopuses masquerading as females don't count is because they are not fooling the *female*, only the male. Now, if a male octopus made himself look like a female to attract a male, well, that would be another matter altogether.

Looking like a female can be one way to get ahead in the mating game. Another beta strategy is to become friends with females. We see the same pattern in humans, which may be why some men are wary when their girlfriend or wife tells them that another heterosexual male is "just a friend." They are right to be suspicious, since male-female friendships in monogamous animals are extremely rare.

In promiscuous species, however, it is far more common. Back to the idea we've already seen with the baboons—friends with benefits. In chacma, yellow, and olive baboon monkeys, a lower ranking male will befriend a female he is interested in, protect her from bullies, and even baby sit the kid she already had with the dominant male. As I've said, females compete for these friendships. With all these benefits, who wouldn't? Often enough the female will come to value this friendship and mate with her buddy next time around.

These friendships can be temporary or long-term, like those seen in the Assamese macaque, endemic to Nepal. Females sometimes prefer these beta males, and they will often willingly consort with them on the side, out of view of the alpha male.[13] I should point out that women ought to be equally concerned about their male's heterosexual female friends. Wild biology is not going to back you up in your argument of "Stop being jealous, we're just friends." From the male or the female perspective, being "just friends" is, most likely, a dangerous tightrope to walk.

In some species, males are so clever that there are three types of males: alpha, beta and gamma. Such is the case with the marine slater, an aquatic isopod that is related to pill bugs, their terrestrial counterparts—you know, those cute little bugs that curl up into a ball? Marine slaters have a traditional alpha male, and like the Indonesian octopus they have the beta-male female impersonator. Then there is the gamma. This male is so small that he invades the harem of the alpha male and succeeds in two-timing both alphas and betas![14]

HARASSMENT AND COERCION—A BIOLOGICAL LOOK

As many women can attest, sexual harassment is a common occurrence. Whether it is at work, at school, at the grocery store, at a club, or even in your own home (the maintenance guy in my apartment complex), it seems that it often comes with the territory of being female. Though women in the United States are much safer than women in many other countries around the world, the incidents of harassment, stalking, and worse are high. Females of all types of species have to deal with this often-constant harassment. As a biologist, I have to look at this phenomenon, not through moral-colored glasses, but as purely a male mating strategy, which across the gamut of animals, including humans, it is.

So, let's look for a moment at this male mating strategy,

which, by the way, often has an unsuccessful outcome—for the male. Female guppies are constantly hassled by belligerent males. One popular tactic among these males is to just thrust themselves at the female without even bothering to court her or ask permission. They just swim after a female, literally trying to stick it to her. This constant unwanted attention interferes with a female's activities and is physiologically stressful.[15]

There are a few ways that the female guppy can deal with these males. One option is simply to move. This choice is not very appealing because she has to move where males are not likely to follow, a location that is riskier. That would be like human females having to move to the North Pole (where we all know polar bears make their home) to avoid aggressive males. A far better strategy for guppies in countering unsolicited advances is for females to hang around a more receptive female, thus redirecting a male's affection. And guppy harassment isn't exclusively for guppy females. No, Trinidadian male guppies have such appalling manners and are so eager to mate that they even harass females of another species.

Eagerness among males leads to all kinds of harassment and even misidentification. The saying "he'll jump on anything that moves" also applies to the world of insects. Many male insects are so lustful that they pounce on the first female they think they see, even if they later discover that the individual is actually a male. That is why up to 85 percent of bugs engage in homosexual behavior.[16]

If you think that harassment by male animals stops with their own species, or even occasionally wrong species because they look alike, think again. Humans don't look much like dogs, dolphins, or sea turtles, yet many a person has been subjected to their unwelcome sexual advances. We can all relate to dogs and the enthusiastic humping of the leg that some dogs are prone to. And why is it always the little ones, like Pomeranians, Shi Tzus, and Westies?

But sea turtles? Why are they molesting people? No one knows for sure yet, but more than one diver has been assaulted by a male sea turtle. Weighing in at well over two hundred pounds, these males try to pin divers to the ocean floor in the typical fashion of mating sea turtles. Very dangerous for the unfortunate diver. One man reported that he tried to give the sea turtle a lobster, only to find that after consuming the lobster the male pursued him even more vigorously.[17] Perhaps the sea turtle perceived the lobster as a "sex gift."

Charlotte, the one who finds bald men super sexy, swears she must give off female dog pheromones because dogs routinely harass her. Like the effect Obsession has on cats, domestic or wild, her pheromone signature makes male dogs go crazy for her. Although she's a self-professed dog lover, she's a little wary of approaching the big ones. She's been leg-humped way more times than she'd like to admit. One time she told me she went into a café to order a café-au-lait (she lives in France) and she bent down to pet a little white puffy dog sitting placidly at his owner's feet. He bared his tiny teeth, growling at her. She got the message and decided it was prudent to respect his space. As she started drinking her coffee, she felt that unmistakable thrusting against her shinbone. No way! Oh yes, and as she reached down to swoosh the humping dog away, he tried to bite her.

Beyond such physical harassment, males can simply intimidate females. It is the equivalent of the man who threatens to prevent your promotion, or worse, threatens to fire you if you do not submit to his advances. Water strider females know all about this male strategy. Water striders are those cool bugs that walk on water. Who knew that their romantic lives are full of strife? In an effort to curb forced copulation, females have developed abdominal spines or shields that can punish an offending male for attempting to mount her. Sadly, while this specialized structure acts as a chastity belt, it does nothing to stop male intimidation.

How exactly does a water strider male intimidate a female into mating with him? It is a risky proposition even for him, but the male will try to attract a predator by tapping on the surface of the water. The female recognizes that the male's behavior will get the attention of a predator and, unable to escape, she chooses the next best thing—to just go ahead, be done with it, and mate with him.[18]

When it comes to humans, men exert power and control over females by sexually harassing them. One could make a similar argument for animals, since in the majority of cases males are faster, bigger, and stronger than females. Whether it is in animals or people, one thing is true for both. The costs to females for this male strategy are high. As I mentioned regarding guppies, females will experience physiological stress, they will move to a riskier neighborhood in order to evade offensive males, and they will lose opportunities to eat by spending their time dealing with these hooligans. Though many human females eventually choose to stand up to their harasser (filing charges, restraining orders, and so on), others can identify with the tactics of guppies; they may resort to leaving their jobs or schools, or even moving away to avoid harassment.

Men get harassed, too, but the frequency is much higher for women. And it doesn't stop with harassment or intimidation. Nearly 17 percent of women will be sexually assaulted at least once in their lifetime as opposed to only about 3 percent of men.[19] Though I have strong feelings about this subject, they will have to wait for another book. For the moment, I will just say that those numbers are way too high for *both* women and men.

In animals, males occasionally use force. When I use the words *force* or *coercion*, I am saying that the female is unwilling or resisting. You may wonder why I don't just say *rape*. It is not because I am trying to sugarcoat the behavior. Quite simply, I am not calling forced copulation in animals rape because, in human language, the word has moral and legal implications that are not relevant to animals, even if the outcome is the same: unwanted, unsolicited, and/or forceful mating.

Male diving beetles fit the bill for a particularly sadistic mating approach. As the name implies, diving beetles dive under water to catch food. Before going under, they collect a few air bubbles to tide them over. Males have suction cups on their front legs, and as a female moves past, the male will grab the female, using these suction cups to hang on. Females try to get away, and in response, males shake them around and then hold them under water. Rather than drown, females give in and mate with the male.[20]

This tactic of the diving beetle is one of the most extreme. In other species, males have come up with other ways to mate with nonreceptive females. In orangutans, it is usually the subadults, or teenage males, that forcibly copulate with females.[21] In little brown bats, getting females while they sleep seems to be a strategy for some males, where even males are victims of sexual assault, by other males, of course.[22]

This might give the impression that forcing an unwilling female to mate is a good strategy. It may be for diving beetles, but in other cases it is usually the *least* desirable males, in other words, males that can't convince any females to mate with them, that use this approach. On top of that, rarely are offspring produced from these couplings. It would seem then that these males are just not the best for some reason, and even through forced means they are rarely successful at mating.

Females are not without options to protect themselves against aggressive males, especially those females that live in groups. In many primate species, females will band together or recruit other males to beat up these aggressive males or to come to the aid of a female in distress. When push comes to shove, a female will also kill outside males in an effort to protect herself.

YOU'RE MINE . . . ALL MINE

Moving away from harassment and force, we still are in the land of less-than-desirable male mating strategies, at least from the perspective of females. What happened to all those gifts and free meat? Mate guarding arises because males do not want females to mate with other males. Females, on the other hand, often want to seek out additional males. This is a conflict to be sure. One effective way for males to ensure females don't produce offspring with other males is to physically keep the females away from other males.

I have already talked about how risky it can be for males to fight other males. Given the size difference between males and females, it is easier and safer for males to control their own females once they have them. Certainly this is true in primates. Hamadryas baboon and chimpanzee males frequently bite, slap, kick, pound, and drag females.[23]

Remember the diving beetle? Well, after a male forces a female to copulate, you would think he would let her go. Absolutely not. It gets much worse. He continues to hold on to her, dunking her underwater, only allowing her to breath from time to time—for up to six hours. This sounds pretty bad, but the poor diving beetle is lucky she's not a walking stick insect. In that species, after copulation, a male may stay attached to his mate for seventy-two days!

When we look at cultures around the globe we see a similar theme: repression and control of women and girls. As humans, we think that this behavior has its roots in religion, culture, or other social practices that many are continuously fighting against, but, once again, my animal studies force me to look at this human social phenomenon through a biological lens. We see that males everywhere attempt to control female movement, access to education, property, or money, to name just a few. In some cultures it is so pervasive that women are even covered up from head to toe. When it comes down to it, this is classic mate guarding.

Human males also engage in other behaviors that are just as extreme. Who do you think invented chastity belts, eunuch harem guards, and a variety of other objectionable practices? Not a woman, that's for sure. If you are a woman on an island in Papua New Guinea your husband might restrict your movement as a means of mate guarding.[24]

In animals, mate guarding is linked very closely with a high probability of reproduction, which means that if the female is fertile, the male is more likely to keep her close. So why do human males do this? Well, for a man in some places, there may in fact be a significant chance that another male *will* father his partner's child. In a village in Trinidad, 16 percent or more of offspring are raised by a male partner other than the biological father.[25] Mate guarding by men might not have evolved if there was not a substantial risk of cuckoldry. This is not the voluntary cuckoldry that is a fetish for some people. In biological terms, being cuckolded means that your mate engaged in copulation with someone else. Naturally, if human males are engaged in mate guarding, and a woman's primary mate does not always father her child, this implies that multiple mating by human females is also common.

JUST PUT A CORK IN IT!

There are less aggressive approaches that males opt for in their efforts to ensure that they father a female's offspring. One rather sacrificial strategy is to break off their penis inside a female. Okay, so we agree that we are all glad to be human at this point. Naturally, when a male donates his penis to the cause, this means that he can't mate again. Perhaps this is why it is not that common? Sometimes this loss is permanent, and other times it is temporary, as in the case of the octopus. The male octopus can regrow his penis, which is called a *hectocotylus*. This is just another trick up the nifty octopus's sleeve.

Perhaps the most extreme example comes from honeybees. Honeybees have a complex social structure called *eusociality*. One of the key features of this type of social system, which is common among social insects, is a division of labor within the group, or colony. This division of labor is a caste system in which different bees assume different jobs. For honeybees, this means there are male drones, nonreproductive female workers, and a queen.

Female honeybee workers have many jobs, including gathering pollen for food, laying unfertilized eggs that will become male drone bees, and regulating the temperature of the hive, particularly where young bees are developing. Male drones are born to reproduce, though only a few lucky ones actually get to mate with a queen bee. Males have no stingers and spend their lives flying around, gathering in areas with other drones, vying for the attention of that one special queen. The nuptials are undertaken in flight, which is a feat in and of itself. These little dudes copulate for ten to thirty minutes. How long a queen mates with a drone is correlated with the amount of sperm he produces. The part that comes straight out of a science fiction novel is that when the male bee ejaculates, his abdomen and sex organ (called an *endophallus*) ruptures, or explodes, killing him.[26] What is the point of all of this? The tip of his endophallus remains in the queen, creating a kind of "plug" meant to guarantee that he will be the only drone to mate with her. Unfortunately for the now-deceased male, the irony is that females can remove his appendage and mate with the next male, and the next, and the next. Well, you get the picture.

Fortunately for males, there are other types of plugs that don't require donating one's penis to what is clearly a lost cause. Plugs—whether you call them *sperm plugs*, *copulatory plugs*, *mating plugs*, *vaginal plugs*, *sement*, or *sphragis* matters not. No, you will not find this type of plug at your local sex shop. A copulatory plug is a gelatinous substance that solidifies upon drying. It is designed to seal off the female's genitals so that other males cannot get in there. Kind of like an animal version of an internal chastity belt.

This is an inexpensive alternative to guarding a female, and it presumably sets the male free to go in search of another female. We see these copulatory plugs in everything from insects to reptiles, rodents, and primates. Now guys, before you think it would be a good idea to plug up your mate like corking a bottle of wine, consider this: other males take those plugs out (of course they do) and females will also remove the plug (naturally).

Case in point: ring-tailed lemurs—you know, those funny looking primates that some people think look like raccoons because of the rings around their tails? Female ring-tailed lemurs are receptive to males for a rather short period of time, like, less than one day. They also prefer to mate with more than one male. With mating going on for such a short period of time, there is a lot of competition among males. So a male will ejaculate into a female and then, while a plug forms, he will move on to find another girl. The problem with this strategy is that the next male to come along just removes the plug placed by the previous male. He then ejaculates and another plug is made.[27] Then, should a third male come along. . . . Easy to put in; easy to take out. Doesn't seem like the plug tactic works terribly well—at least not for the male.

In tree squirrels, males don't remove the plugs placed by other males. Instead, it is the female that doesn't take kindly to being plugged up and having her options limited. Immediately after copulation, females groom their genitals and use their teeth to pull out the plug that was so rudely put in place. What does she do with it? Either she eats it or she just throws it away.[28] So, it would seem that to some it's a yummy treat and to others it's just another chintzy gift.

THE HARD FACTS ABOUT SPERM

If these plugs just get removed, what is the point? In the end it may just be to give one male's sperm a bit of a time advantage, a head start, if you will. Ultimately, sperm plugs are one way to help sperm compete. Sperm competition is only necessary if females mate with multiple males, meaning that a single male's sperm doesn't compete in a vacuum. We already know sperm can hook up to become bigger or longer, thus providing an advantage. Disproportionately large testes (relative to body size) are also an indication of sperm competition. And, as we now know, the human penis acts as a shovel to get rid of the sperm from other males. Human males also thrust faster if they have been away from their partner for a period of time. Where there is promiscuity, there is a need for sperm competition as a strategy.

Overall, sperm appears to be rather innocuous. After all, its role is simply to fertilize the egg. But not all sperm likes to compete, and plugs aren't for everyone. So what's a sperm to do? Manipulate your girl with chemicals and hormones. In other words, use the seminal fluid that transports your sperm to make voodoo on the female.

You probably think I am exaggerating, but I am not. In fruit flies male seminal fluid contains all kinds of chemicals that make the female permanently unattractive to other males.[29] In addition, these chemicals are downright toxic and can shorten a female's life span.

What about in humans? Many a woman has confessed that she fell in "love" with her mate once they'd had unprotected sex. Is there any truth to that? Does semen affect our behavior and physiology? Given that human male ejaculate contains a plethora of chemicals and hormones, mostly antidepressants, it seems that exposure to these substances could then very well have a positive effect on the mood of the recipient.

We still don't have definitive results on whether male seminal fluid can manipulate women into being faithful, but it is clear that women who have unprotected sex (or consume semen) are less depressed than their safety-conscious counterparts.[30] This begs the questions: Is it true love or are we just happier because we are exposed to all the good stuff found in seminal fluid?

THE CHOOSY BACHELOR

With all these strategies that males have, it might give the impression that males are just out there indiscriminately attempting to mate with anything that moves. Okay, maybe that is true for bugs (and guppies). A good part of the reason for this is that it is usually the female that is making a larger investment. She is producing the egg, which is larger and more energetically costly than sperm, plus she (except in rare instances) is carrying the offspring in one form or another, and, in many cases she is raising the offspring with little input from the male. It makes perfect sense then that males would compete for females, and that females would be choosier.

So what makes some males very picky? The game changer is when both the male and female are required to raise the offspring, as in monogamous species, or when the male is also making a high investment. This investment need not be helping to raise the kids. It could be something like donating his house.

This is the case with the nocturnal sand-dwelling wolf spider. Males are larger than females, which is rare in spiders. These spiders are homebodies, living in burrows sealed with silk and sand. Females get around a bit more while looking for and courting males. They open the door to a male's burrow and wave their legs around. This really turns male sand-dwelling wolf spiders on, and their bodies begin to shake in response to these leggy ladies.

The female then follows the male into his man cave. Once deep inside the burrow, if he likes her, he switches positions with her. If he doesn't switch positions, it means that she didn't make the cut. She'll turns around and head back out the way she came in. But if all goes well and they mate, the male generously donates his home to her, and both of them cooperatively seal off the entrance.[31] This may not seem like such a big deal, but think about your own man cave, gentlemen, and consider whether you would willingly *and* happily vacate it for the exclusive use of your lady. Permanently.

For the male sand-dwelling wolf spider there is a double cost incurred as a result of this generosity. Not only has he lost his home, which means he'll have to expend energy to build a new one, but it is also risky to be running around the sand dunes without a safe burrow to dart into.

Perhaps male seahorses are the choosiest. After all, they bear the brunt of all the domestic duties. And I do mean all, including getting pregnant and providing parental care to all the baby seahorses. Huh? Pregnant males? Yes. The female deposits her eggs into an enclosed sac on the male's tail, after which he fertilizes them. Once that happens, the developing embryos implant into the pouch wall, where they get air and nutrients until, *voilà*, he gives birth to baby seahorses.[32] Baby seahorses are deliciously cute, aren't they? In the pot-bellied seahorse larger females produce more eggs, larger eggs, and subsequently, larger babies. Just like females prefer larger males when it is the moms doing all the work, male seahorses have a clear preference for larger females.

In many species where both males and females provide parental care, both sexes will tend to be choosy, but even in cases where mating is promiscuous males show clear preferences. For example, mandrills, which are closely related to baboons, lives in enormous groups, and the males are spectacularly colorful. Females have a colored snout as well, but nothing compared to the males.

Because males are larger and more colorful, conventional wisdom would predict that they are not that choosy when it comes to mating. After mating with females, male mandrills guard females to prevent access by other males, but they seem to preferentially guard high-ranking females. Low-ranking females? Eh, not so much.[33] Apparently, males care more about whether another male mates with their female if she is highly desirable.

Why are high-ranking females sought after? High rank equals high quality in mandrill society, and the offspring of high-ranking females are more likely to survive. Given that the dominant male gets to mate with most females, and that many females are simultaneously available, the dominant male hardly has time to guard them all. The solution? Spend his time and energy guarding only the best of the best.

This preference for high-ranking females is one of the primary reasons male chimpanzees prefer older females. We saw that there were benefits for older females that mate with younger males. Here, we can see that there is a clear benefit for males that prefer older females as well. What is the benefit in chimpanzees? First, simple experience. Young female chimps are inexperienced, especially when they get their first sexual swelling. They may be afraid of males, running from them or otherwise trying to elude them. They are also not very experienced moms, so survival rates are higher for offspring born to veteran and high-ranking moms.[34] Another benefit is that rank passes down through the mother in chimpanzee society. Queens give rise to kings, and all males want to be kings or to have their sons be kings.

From bumblebees to spiders, birds, and fish, another way for males to exert their choosiness is to do it while they are mating with a female. Just as females in some species will mate with multiple males then sequester sperm, kill sperm, or allocate sperm preferentially, males may adjust the amount of sperm they give a girl, giving more or less based on her "quality."[35] Quality can be determined by any of a number of variables, including whether there is sperm competition, whether the female has already mated, or even the extent of the female's genetic compatibility. It is kind of like the guy who thinks a woman is good enough to have sex with or even date, but definitely not good enough to marry.

WHEN THINGS GO TERRIBLY WRONG: THE MALADAPTED MALE

This was far from an everyday occurrence, but one day when I was leaving the grocery store I saw a squirrel in the parking lot. As I am apt to do, I started having a conversation with the squirrel. Adjacent to me, about six feet away, was a man, and he made some remark about me talking to the squirrel. Aside from a quick glance in his direction, I was so engrossed in this one-sided conversation that I hardly paid him any attention. After saying goodbye to my squirrel friend, I happily went on my way and arrived home with my groceries.

As I headed up to my apartment on the second floor I heard a voice call out to me from below. When I peered over the railing, there was the man from the parking lot. He said, "I hope you don't think I'm crazy, but I was just wondering if you would like to go out with me?" He could have been completely harmless, but any stranger who follows a woman home puts her in a dangerous position, so I dealt with him cautiously.

I could have refused his request for a date, but I decided it was smarter to say, "Hold on a second. There is someone I would like you to meet." I went inside and came out with my 210-pound Great Dane, who, right on cue, began snarling, growling, and barking. I could always count on him to protect me. The man scrambled into his car with a terrified look on his face. I was pretty sure he wouldn't be coming back looking for me. Although women also stalk men, there is one big difference in the pattern. Men will stalk women they do not know. They will stalk complete strangers.

But women, especially in younger age groups, do physically assault men more frequently. The difference is, when a male attacks, he often causes serious injury.[36] Domestic violence in human relationships crosses all social boundaries—no one is safer than anyone else due to socioeconomic status, color, or sexual orientation. As we have seen, males of many species are aggressive toward females during mating or to make sure the female does not mate with other males. During the day, olive baboon females are subjected to male aggression, and 25 percent of these aggressive encounters result in a physical attack.[37]

Why so much aggression? Though we've looked at aggression linked to mating issues, in these baboons a large percentage of the aggression is related to food rather than to sex or sexual jealousy. Males are simply stronger and bully females to get what they want. On the other hand, the silverback gorilla is routinely aggressive toward his females, but more often than not, his aggression is used to break up fights among the females.

When it comes to long-term partnerships, like lifelong monogamous pairing, I cannot think of a single example that is characterized by the systematic abuse (verbal, emotional, and/or physical) that we see in humans. On the contrary, aggression appears to be reserved for outsiders and other potential dangers. The monogamous prairie vole comes to mind. When a male and female meet and decide to pair up, there will be a whole lot of lovin' going on, from grooming to cuddling and everything in between. Once the pair is bonded, the male becomes extremely aggressive, attacking anyone, male or female, *that is not his partner*.[38] He definitely engages in mate guarding, but not by attacking his mate. If a prairie vole male were to attack his mate, bond or no bond, I suspect her response would simply be to try to get away.

Although it is not easy to get away, female animals generally *do* try to escape abusive and violent males. After all, sometimes males can go too far, and the female can be injured or killed. The sex ratio in a given population is usually equal or skewed in favor of females. Strange things happen when this balance is disrupted. On a French island called Île Longue this trend is reversed; there seem to be more male than female sheep.[39] What is causing the premature death of females? Males. And it is terribly brutal. The mating system on this island is promiscuous, so both males and females mate with multiple partners. During the rutting season, males completely lose control of themselves. Multiple males will relentlessly chase a female and attempt to mount her if she stops. The males, all vying for the mating opportunity, will ram each other and inadvertently her, causing injuries to the female. Frequently, the females die as a result of the injuries suffered during these melees.

What is causing such violent maladaptive behavior? There is no leader among the males. It is anarchy. Without an alpha, the males are unchecked, running around like a gang of vicious hoodlums.

Similar things can happen with lizards if you manipulate the sex ratio and artificially create more males than females in a population. When there are too many males running around trying to mate, instances of female injury and death during mating increase.[40] I suspect that given the very large sex-ratio biases we now see emerging in India and China as a consequence of the social practice of killing infant girls, we will see increases in male sexual violence toward females.

The main difference I see between humans and animals in this regard is that injuries to female animals that frequently occur during mating are accidental. Meaning, unlike in humans, where rape is about control, power, and aggression, in animals, males are not intentionally trying to hurt the females. Furthermore, the behaviors we discussed earlier in lizards, sheep, and even ducks do not represent the norm.

In humans, we see domestic violence manifest itself usually as a consequence of sexual jealousy. The higher the value of a mate, the more intense the guarding (jealousy) and the greater likelihood a man will resist the relationship ending (often by engaging in stalking and violence). As we saw with women, a percentage of men fall into the category of "maladapted." The aggressive behaviors of such males do not lead to creation of a cooperative and positive coupling, and this reduces success when it comes to raising offspring.

Although I started this chapter suggesting that we take a moment to appreciate how difficult and even dangerous it is to be male, there are some downright dangerous males out there. If we take our cue from animals, we can see that trying to figure out *why* a male is acting threatening is not nearly as important as recognizing *when* a male is a hazard.

At some level, we have been conditioned to be more accepting of male aggression. But we can't just shrug our shoulders and say, "Oh well, it must be natural." Clearly not all males use these behaviors as mating strategies. Knowing what to look for and what to avoid can help keep a woman safe. I think it is time we abandon the phrase "boys will be boys" and start changing our ideas about violence.

At the same time, thanks to Oscar (and a few other male friends), I came to see that males face a different set of challenges, which women will probably never be able to completely understand. Maintaining one's alpha position, giving the right gift, sperm competition, guarding one's mate, the Napoleon complex—it's enough to make anyone feel a little overwhelmed. In addition to the biological hurdles of finding a mate, men, like women, also have to fit a socially accepted image of what it means to "be a man." The bottom line is that it's a tough world out there for males.

7

ARE WE MATING OR DATING?

One night, after lamenting endlessly the state of my romantic affairs, yet again talking with Oscar, he asked me point blank, "What is it you want? What are you looking for?" I stopped speaking, opened my mouth to say something, closed it, and opened it again, only to utter a stumped "uhhhhh. . . ."

Through the fog of wine I suddenly had a clear vision of what I wanted. I told him, "I want someone who will be my friend," then I continued, saying, "someone to be my partner, who will have my back, not stab me in it." I was on a roll. "I want to be able to do things together, cheer each other on, and, of course, I want to be attracted to him." He simply replied, "Very well then. Are you *acting* like that is what you want?" *Whoa . . . what?* Then it occurred to me that we humans sure do complicate things. As years of research papers flashed before me, I realized that we, like our animal friends (both male and female), have a multitude of mating strategies, and maybe sometimes we don't know which ones we are using.

If I were a thirteen-lined ground squirrel my romantic life would be a lot less muddled. For these squirrels, both male and female, it is pretty straightforward: a quick hook up is the only way to go.[1] Males make plugs, there is sperm competition, males and females mate multiply, and once they mate, the females raise the litters, and that's all she wrote. The female and male are not confused. They don't have to think that now, just because they've mated, they are making their way down the aisle with the eyebrow raising baby bump and the "holy crap, what are we doing?" look on their faces, joined at the hip for all eternity, destined to find, hide, and share their nuts together—forever.

We can think of the quick hook up, as practiced by thirteen-lined grounds squirrels, as a short-term mating strategy. Meanwhile, the lengthy partnerships of gibbons, geese, and penguins are models of the long-term-mating strategy. But are things really this simple? Unlike thirteen-lined grounds squirrels, which only pursue a short-term approach, other species, including humans, combine short- and long-term tactics depending on their agendas; often, individuals pursue multiple agendas simultaneously.

Let's put it this way, while thirteen-lined grounds squirrels get together for a minute, we can get together for twenty minutes (okay, maybe seven) or for a night, a weekend rendezvous, a summer fling, a season, a year, or a lifetime. Among humans, this is truly universal, with different cultures having different words for these various types of interactions. Despite all evidence to the contrary, the strong cultural belief that humans, as well as other "monogamous" species, pursue only one mating strategy continues to persist.

There are a variety of ways we can mix and match short- and long-term strategies. A man named Griffin was a great example of this. Somehow, as soon as I met him, I knew I would never forget him. Of course my reasons for remembering him now are far different from what I thought they would be at the time. I met him while I was dancing the night away at a club in Fort Lauderdale years ago.

In a classic "our eyes met from across the room" scenario, our eyes literally did meet across the dance floor. My eyes followed him as he made his way around to where I was. Van Morrison's "Brown Eyed Girl" was playing, and somehow I just knew that he was the guy for me. Soul mates! Silly girl. We danced all night, exchanged numbers, and then, the following day, he took me to lunch at a posh Mediterranean place right on the water. We talked and walked, and then at the end of the first date, we ended up at his place—the yacht.

Now you may be thinking that's all there was to it. But no, at the time he was a private chef on a yacht and the ship was docked in Fort Lauderdale for a few weeks. I was much younger and not as wise back then, and what I thought was a whirlwind romance that would last a lifetime turned out to be merely something to occupy his time while he was in town. Perhaps I got confused because he showered me with attention, dates, talks of love, and romantic evenings for several weeks.

These things, I thought, went outside the bounds of short-term mating strategies. Not so. The ship left port, and with it went his affections. I never heard from him again. I did find out the following year, when the boat returned sans chef, that he'd had a serious love interest back home the whole time, and that he was due to be married. Some of his friends felt badly about what he had done; they told me everything. Clearly he was pursuing a long-term strategy with his other and simultaneously practicing a short-term tactic with me (and likely many more). No wonder it can get confusing out there!

The reality is that both males and females employ a variety of mating strategies depending on whether they are pursuing a short-term or a long-term partnership. There are costs and benefits to each, and maybe the only problem is when you don't know what you are doing or you get hoodwinked or confused about what someone else is doing.

BENEFITS OF A ONE-NIGHTER

To begin this discussion, one really must set aside social stigmas about females versus males and one-night stands or short-term flings, since statistics clearly show that both men and women partake in the "no strings attached" roll in the hay. Instead, it is more interesting to focus on why we even have one-time or brief encounters. Is there any benefit? And do we behave differently if this is what we are up to?

Let's distinguish between one-night stands and other short-term mating opportunities. We can think of the classic one-night stand as a single encounter with a stranger. Don't worry; we'll get to hook ups, booty calls, vacation romances, and friends with benefits a little later. The number of species in which individuals come together, mate, and then part company is too long to list, but there are huge reproductive advantages for both males and females. Whether it is the fly-by-night fertilization of fishes or the brief encounters of Tasmanian devils, promiscuity often leads to greater reproductive success.

The key is that, more often than not, females (we're talking species wide) that mate with multiple partners often increase their chances of getting pregnant, and they produce a higher number of offspring. The common thread, in short-term mating strategies, is that there is no expectation or need for the male and female to remain together after mating occurs. Furthermore, with such fleeting episodes based exclusively on sex, there is little opportunity to assess an individual beyond his or her physical features and/or displays and, therefore, genetic quality. No need to figure out if the fly-by-night hook up will be a good provider, partner, or parent.

It is clear that we have definite preferences for certain features and traits, but do these preferences change depending on the type of mating strategy we are pursuing? When it comes to women, the primary concern for short-term mating, like the one-night stand, is physical attractiveness.[2] No one is doing much assessing of the male's long-term mating potential, ability to provide (resources), or even intelligence. Nope, women don't even care whether the males are particularly nice. The proof is in the pudding, and this pudding is judged pretty much on how "hot" it is.

The results vary somewhat between men and women, with women being more purist in their pursuit of the "killer hot" one-night stand. What this boils down to is that highly attractive men are (as we all suspected) much more likely to add notches to their one-night-stand belt than their "average" looking rivals. When it comes to one-nighters for men, studies show that they tend to be a tad more flexible when it comes to attractiveness. Although attractiveness is still considered an important quality, men are many times more likely to "settle" for a less attractive partner who is willing to engage in a one-night stand.[3]

I remember seeing a documentary about this several years ago. A young woman and young man were sent to a college campus in the United States. Their job was very simple: they had to go up to members of the opposite sex, tell them that they had seen them on campus and thought they were attractive, and then ask them if they wanted to go back to their dorm room to have sex. Both the woman and man were relatively attractive. Any ideas what happened? Well, not one woman said yes, and most of the women were outright offended. On the other hand, only one man said no, and he apologized, explaining that he had a huge exam and needed to study. For men, there doesn't seem to be any reason to turn down a sexually willing female. Though responses were very different depending on the sex of the candidate, I would be interested in seeing the study repeated with an extremely attractive young man.

The findings regarding hook ups are likely also true of other short-term mating strategies, such as sex vacations, booty calls, and friends-with-benefits arrangements. I do feel there is an important distinction that must be made between the booty call and friends with benefits. Case in point: Steve. I had seen him around for about two years, and we always said hello, offering a smile and a wave to each other as we passed. Since I don't walk around expecting to get asked out, I was pleasantly surprised when one day, out of the blue, he asked if we could get together.

We exchanged numbers, and I was happy to have been asked out on a proper date. We met for lunch and talked easily. He was good-natured, funny, and intelligent. And then he said, "Well, you know I recently got separated, and I am not really looking for a relationship." To which I replied, "What are you looking for?" He said, "A friend." Oh, boy. I smiled sweetly and said sincerely, "I can totally understand that, and you should definitely take your time and have some fun before getting into another relationship. Unfortunately, I am looking to date seriously and am not interested in what you are looking for. However, I would love to get to know you as a person, become actual friends, and have lunch or coffee with you anytime."

A hug and a goodbye after lunch, and that was that. Or so I thought. He continued calling me, and he talked about all the places he would like to take me. His suggestions included week-ends away in the mountains, Hawaii, and other extravagant destinations. My reply was always the same, "Coffee or lunch would be a fine way to get to know each other and become friends." Then one day he said we were already friends and wasn't it time we enjoyed the benefits. I laughed and pointed out that friends spend time together and do things together.

Perhaps because I am an animal behaviorist I can laugh at these antics rather than be offended, but I did stop taking his calls. The point I am trying to make is that a friend with benefits is actually a friend first and then (depending on both parties and the situation) there may (or may not ever) be sexual perks to the friendship. Remember the baboon males who were "friends" with females? They were truly friends. The male would help the female, groom her, protected her, and baby sit for her. *Friend.* So fellas (and ladies), please refrain from suggesting you are looking for a friend when all you want is a booty call. Nothing wrong with either one, but they are different, and a lot of time and resources can be wasted in the pursuit of one if you are not clear about which you are actually seeking.

Obviously Steve and I had different agendas, but if both parties are happy with whatever arrangement they have agreed upon, that is fantastic. Problems arise when an arrangement has been struck, and then one party tries to change things. That would be like a male thirteen-lined ground squirrel proposing to a female he just mated with! Awkward. Let's say one person was under the impression that a booty-call arrangement had been struck, but now the other person wants to be friends and hang out. Worse, maybe one person now wants to start dating seriously.

Quite a few years ago, a friend of mine, Julia, came to me asking for advice about a guy. I happened to know him, too, and had never suspected there was anything going on between them, which made giving my input a bit strange for me. Julia explained the particulars of their sex-buddies arrangement, not quite booty call and not quite friends with benefits. This "buddy" would visit her once a week, and they would spend the night together. Within fifteen minutes of his arrival they would be in the bedroom. She said it had been going on like this for months without a glitch. In the morning he would get dressed, give her a quick smooch, saying "See you next week!" She said that they never spoke during the week, even though they worked in the same building, and their only other communication was text messaging to confirm or change plans, if necessary.

One night, after several months, he came over earlier than usual and asked about her week. Unsure of how to respond, she kept the conversation brief. It seemed uncomfortable for both of them, and after a few minutes they gave up and went on to business as usual. Then, a few days later, he called her during the day, saying he just wanted to talk. Shortly after that he suggested dinner before settling in for the night, followed by breakfast in the morning. She confided that she really didn't want to change their agreement. She was just recently out of a long, difficult relationship, and she was happy with things the way they were. On the other hand, she didn't want to hurt his feelings or create a strange workplace environment.

I told her that since they were upfront with one another at the beginning, it was fair she should tell him how she felt. Shortly thereafter, she ended their weekly trysts. I don't think anyone we worked with ever knew, but I have to say that I did notice an ever-so-slight chill in the air for a few months after her decision to call things off.

DOES IT PAY TO PLAY?

When it comes to short-term mating strategies, the benefit for animals is obvious. There are significant reproductive advantages, and this is reflected by the fact that these decisions are still made in accordance with biological principles that will produce the best outcome in terms of offspring. Of course animals don't "know" this; they just do it for the same reason we do. Like humans, they have a natural sex drive because sex is fun and it feels good. More on that later. For now, given that most of us actively, consciously, and deliberately avoid successful reproduction in short-term mating situations, and both women and men alter their selectivity, what are we getting out if it?

There is some debate, especially about what women are getting out of short-term encounters. There are many researches out there today who say that females engage in short-term mating strategies because (a) they get access to good genes, (b) it promotes sperm competition, (c) they want to confuse paternity, receive protection, and gain additional resources, and (d) it affords them opportunities to evaluate men for their long-term potential.[4] As I have stated before, I disagree with all but the possibility of gaining access to good genes, though this benefit is rarely realized. Certainly, a one-night stand does not provide protection or resources, and there is no real opportunity to evaluate the male in a single encounter. Because women report being less sexually satisfied than men in these situations,[5] it seems unlikely that women are able to accurately assess a man's sexual prowess.

Other researchers look at short-term strategies from more of a psychological perspective, believing that the benefits may include things like feeling desirable, feeling socially connected, and having sexual variety; according to some women, short-term strategies also provide help with getting over one man by getting under another. Women who are socially isolated find short-term mating strategies more appealing.[6] Though the ideas about male promiscuity are also evolving and changing, and one cannot exclude the possibility that there is a psychological component to male short-term mating behavior, the general consensus is that for men (especially young men) the motivation is more clear-cut: sex, sexual variety, and sexual pleasure, though my guess is that males, too, find hook ups less satisfying. In my opinion, these "benefits" also apply to women, although the pleasure part may only apply in situations like booty calls, sex buddies, or friends with benefits as a result of increasing comfort with the partner.

Maybe there are true benefits, and maybe there aren't, but are there any costs to a short-term mating strategy? When we look at animals we see that rarely in nature can you truly have your cake and eat it, too, meaning that there are costs to every strategy. What about animals like the thirteen-lined ground squirrel? What are their costs? The primary hazard to promiscuous animals is, just as in people, an increased risk of sexually transmitted diseases. That's right, animals get STDs, too.

The two-spot ladybug gets its own version of crabs with a sexually transmitted mite. Unlike a human case of the "crabs," the mite is anything but harmless to ladybugs. The mite causes infertility in females and increased male mortality over the winter.[7] Just like in us, not all STDs kill the infected lady bugs or make them look sick, which means the pathogen can remain in circulation.

This could be an argument for monogamy, yet regardless of the presence of STDs, promiscuity reigns. Why? As explained earlier in this chapter, females that mate with multiple partners often increase their chances of getting pregnant, and they produce a higher number of offspring. Therefore, this benefit offsets the cost of STDs. Maybe that was the case for us at one time, but with the advent of birth control and the option of having fewer or no offspring, this benefit does not seem to have the potential to offset the cost in humans.

The ultimate cost of one-night stands is most keenly felt in those animals that pay the highest price—death. Remember the unfortunate bee whose endophallus exploded after mating?

There are others, too, that live just long enough to reproduce and then die, though admittedly in not nearly as dramatic a fashion. Brazilian slender opossums, antechinuses (little marsupials that look sort of like hedgehogs), salmon, squids, some spiders, a few amphibians, and a bunch of insects join ranks with the bees. However, the mayfly may just take the cake. If they are lucky, they have a few hours (some species of mayfly only thirty minutes) from birth to death to somehow find a mate! But in all fairness, these are extreme cases. Let's head back to the welcome confusion of having to figure out if the benefits are worth the price one pays.

For women there can be a suite of additional costs. Beyond unplanned pregnancy, a woman can suffer a sullied reputation, lowered value as a long-term partner, and potential violence at the hands of her mate (if she is cheating).[8] Most of these social costs are imposed on women not just by men, but also by other women. Is this why males engage in short-term mating situations more frequently or at least more openly? Or maybe they don't. Remember, promiscuous men have to be having sex with someone. So perhaps women are simply less likely to confess how often they want to (or do) engage in casual sex. I'm not sure that this has been settled, but there is research that indicates males overstate their sexual conquests whereas female underreport theirs.[9]

Another potential pitfall that seems to be common for both men and women is having feelings of regret the morning after a casual hook up. Of course alcohol and drugs can distort anyone's perception, and these may play a role in the willingness of two individuals to pursue a short-term mating opportunity. But because men are much more apt to lower their attractiveness threshold, their chief complaint is that their partner wasn't attractive enough.[10] For women, since their choices are already primarily based on attractiveness, their regret centers around the feeling that they have been deceived. The problem may stem, in part, from the fact that men will occasionally drastically lower their standards, and many women are taken in by the flattery. But once the night is over and the flattery is turned off, they end up feeling used and even worse about themselves. On the other hand, a woman may feel regret because casual sex is substantially less physically gratifying for women than men. Women are also more likely than men to mistakenly believe or hope that something more substantial will come of any sexual encounter.

And with this point, we arrive at where humans diverge from most of their animal cousins. Let's call it misunderstanding the basic nature of a one-time mating opportunity; which is to say, mistakenly trying to turn it into a long-term relationship. This can be an unexpected cost, but it shouldn't be—unexpected, that is. It is particularly bad for one who feels used and rejected because he or she had hoped that the encounter would turn into something more. But think about it. Two people see each other from across the bar, strike up a conversation, and the guy is good looking and the woman is willing. *Voilà!* Both are poised to have a one-night stand. And that is really it.

The same is also true of most other short-term mating approaches. Why? Again, largely because we make our decision to enter into these situations for reasons other than the presence of traits we actually desire in a long-term mate.

When we engage in casual sex, we are not evaluating men or women on traits that indicate long-term potential. It would seem, from a biology perspective, that it makes no sense to suddenly consider our casual-sex partner a great candidate for a long-term relationship based on short-term sexual encounters, whether singular or multiple. A colleague recently asked me if any other animal uses short-term strategies to "hook" a mate into a long-term relationship. The answer is that, with few exceptions, they don't. And when humans try to do this, the benefits will very rarely outweigh the costs.

That is not to say there is no infidelity among animals that form long-term partnerships. The difference is that usually (though not always) the animal is not engaging in a short-term strategy with hopes of transforming the single sexual encounter into a long-term partnership. Rather, the animal is already in a long-term partnership and then happens to engage in extrapair copulation.

In some birds, for example, short-term sexual encounters occur when two birds arrive at the nesting site before their long-term mates. While awaiting the arrival of the long-term mates, sometimes sex with a willing participant just kind of happens. Oops! But the difference is that when the partners arrive, the long-term couples are reunited, and it is on to raising the family as usual. Well, except the offspring may not belong to the late-arriving male! At any rate, the affair is not intended to create a new long-term partnership in place of the existing partnerships.

This seems contradictory to the strategy females use to trade up. From a biological perspective, I would predict that unless a woman looks at traits that indicate long-term potential, the new relationship will fail. If you are a woman, all you "know" at the point of union is that, based on physical attractiveness, this person is likely a good genetic match with whom to produce offspring. For men, she may or may not be physically attractive, but she is willing. But neither party has much information regarding long-term compatibility.

There are exceptions, and sometimes, like we see in humans, animals in monogamous relationships have extended "extramarital affairs" that result in mate switching, meaning that a short-term mating strategy, if it goes on long enough, can be a way to assess a new partner and make a change.

But in the end, short-term strategies can work for both men and women if they are both honest with one another (meaning no masquerading as a long-time lover if you're looking for a one-shot fling) and with themselves. Like the question posed by

Oscar that left me stumped for a moment ("What do I really want?"), it is important for each party to consider exactly what she or he wants out of any interaction and why.

If a wham-bam-thankya-hot-man is what a woman is looking for, then the choices are pretty limitless. But if a woman is having sex because of the few hours of flattery that preceded it, or as a way to feel better about herself and find a long-term partner, then it's probably better to try some more appropriate strategies for finding a long-term mate. Same thing goes for males.

If you are interested in serious dating, there is no getting swept away and committing too quickly. First, you have to be sure that both parties are *truly* interested in a long-term relationship. Since this is not always the case, at the beginning of a relationship it is always best to make sure that both parties are practicing similar strategies.

Some people claim they are "casually" dating. In my opinion, casual dating is really just casual sex in disguise. One is utilizing either a short-term strategy or a long-term one, or pursuing both strategies with different people. And since the two strategies require different sets of criteria, trying to pull off both simultaneously with the same partner is next to impossible. Once again, from my perspective, it is better not to blur the lines between the two, since doing so almost always leads to utter confusion and delusion for everyone involved. With that said, let's move on.

DATING AS INFORMATION GATHERING AND JUMPING HURDLES

Determining that both parties are serious about looking for a relationship is the first hurdle. The bad news is that only over time can one tell whether the relationship has a chance of success. And as for those hurdles? Well, they only get harder.

A lot harder. But before we get to that, let's say for now that you have determined that both you and your potential mate are pursuing a long-term strategy. You already know you are attracted to him or her. What's next?

Complex, extended courtships, which we see in animals, are designed to provide each partner with enough information to determine whether the match will be a good one. This is especially true in those species that are socially and/or genetically monogamous, and when both the male and female are needed to successfully raise the offspring. By necessity, then, the courtship rituals are more involved so that the individuals of both sexes can assess the other carefully. In a sense, we can think of serious dating in humans as an opportunity to gather information. In other words, think like an albatross.

Albatrosses are remarkable. When I traveled to New Zealand some years ago, even knowing ahead of time what wildlife I would find there, I was still amazed by the plethora of "strange" animals. I was in biologist ecstasy. One stop was at the Royal Albatross Centre in Dunedin. From a foggy glass window, neck craned, I managed to catch a glimpse of the royal albatross. Its wingspan can reach eleven feet or more, allowing it to cruise through the air without wasting its energy flapping. Helpful when you may fly over two thousand four hundred miles for a meal.

All albatrosses are long-lived, take many years to reach sexual maturity, and have elaborate courtship rituals. Like the royal albatross, the wandering albatross has one of the longest, most elaborate courtship periods documented, second perhaps only to humans. Young albatrosses start looking for a partner when they are sexually immature; preteens, if you will. They dance; they yammer; they sky call to each other. And this, over the course of several years, is a way to practice. From one year to the next, young albatrosses perform these displays to fewer and fewer individuals, until finally that special one is chosen.

From that point on, the pair establishes a bond, spending less time displaying and more time yapping. A pair may even develop its own "language."[11] No one really knows exactly what the albatrosses are talking about, but maybe they are calling each other cutesy nicknames like Snookums, Boo, or Shorty?

Regardless, they endure this long process to ensure compatibility. Why all the effort? Raising a baby albatross is a long and arduous process that takes a year. If the parents fail to raise a chick to adulthood, they will not try again until the following year. Both males and females share foraging and feeding responsibilities, and, like the penguins, they rely heavily on one another. If one is an unreliable mate, the consequences are severe.

You've probably had the experience of being in a dating relationship where everything is going amazingly, and then, somewhere along the way, usually between three and six months, things change. Turns out you weren't dating who you thought you were dating; you had met that person's alter ego, the person they thought you would be interested in. Perhaps the most time-saving dating behavior human beings could engage in would be to just be oneself—from the outset.

I dated one such person, briefly. He was adamantly and unapologetically himself (as I always try to be as well). The benefit of this approach was that we were able to rapidly determine that we were not suited for each other, freeing us both to choose again. Not unlike barnacle geese.

Barnacle geese breed in the high arctic, building their nest high on the cliffs. As an aside, in the 1500s and early 1600s it was commonly believed that fully grown barnacle geese emerged from barnacles. At the time, it was the most logical explanation for the sudden appearance of large flocks of these geese in Scotland and England.[12] Obviously people were unaware that their disappearance and reappearance was the result of migratory behavior. Even though it was figured out that geese don't magically appear from barnacles, the name stuck.

When they are about one year old, young barnacle geese start "dating." Like humans, albatrosses, and other species that form long-term, sometimes lifelong partnerships, dating is critical to barnacle geese. The point of dating is to sample mates—you know, play the field. What do these dates look like? A pair of barnacle geese that might be interested in each other is together constantly. But these are often temporary relationships during which they are giving partnership a trial run.[13]

Not unlike a couple I knew years ago, Sharon and Randy, who went on a three-month hike, alone, along the Pacific Crest Trail before deciding to get married. Granted, they had already dated for a while and were living together, but they figured being together twenty-four/seven in the woods for three months would really test the waters.

In barnacle geese, these trial runs last anywhere from a few days to a few weeks. Some are lucky and find a partner right off the bat, and others have as many as four relationships before settling down. Wouldn't you like to be a barnacle on the wall to see what went wrong in the brief liaisons? Unlike people, who continuously break up only to make up and break up again, despite the fact that the majority of these relationships do not work out the second time around, barnacle geese accept that a partner isn't right for them and move on, never "going out" with that partner again.

What kind of information is useful in determining long-term compatibility? It depends on the species. One thing is for certain; for many, finding someone who is similar to you is the way to go. While the idea that "opposites" attract pervades our culture, it is not a commonly applied principle in animals when it comes to long-term relationships.

Setting aside homosexual pairings, what kinds of similarities are important? If you are a great tit bird, your sense of adventure or lack thereof is pretty significant. Great tits are an insectivorous little bird that is common throughout Europe, the

Middle East, and parts of Asia. For simplicity, you could divide great tits into two baskets when it comes to the continuum of personality types: those that are homebodies and those that are explorers. These differences in personality have a profound effect on how these little birds cope with life. Explorers are more aggressive and bold, and they are relatively unfazed by changes to their environment. Homebodies are slow to explore, pretty easygoing, and in tune with their surroundings. If you put two people together who differed so dramatically in how they approach life lived together, chances are there might be a lot of conflict in the relationship. The same is true for great tits. Partnerships made up of great tits with similar personalities are more successful.[14]

If you were beginning to suspect that similarity between human partners—on personality, behavior, intelligence, religion, and other values—is beneficial for relationships, you would be correct. This is because similarity in these variables can contribute to overall compatibility. Relationships involving partners with similar interests and life philosophies, regardless of sexual orientation, have a much better chance of survival over the long haul. Individuals with different personalities, behaviors, and values will also often have opposing agendas. How can two people coordinate their actions successfully when they are trying to accomplish different things? This is why most online dating and matchmaking services pair individuals on the basis of similarity.[15]

One of my good friends found someone she believed to be her perfect match through an online dating site. She thoughtfully and accurately filled in all her information, describing herself and exactly what she wanted in a mate. The dating service gave her more than a few options, and she chose one. She was somewhat skeptical, but after many months of "information gathering" she decided he was in fact right for her, and a year after their first date, *she* popped the big question. For all you romantics out there, he said yes and they just bought a bigger house to accommodate the dogs they don't have but both want (common agenda!).

The challenge, however, is that people may be dishonest about where they stand on a variety of issues and behavioral traits. For example, if you value monogamy and your partner values your monogamy but not his or her own and fails to inform you of this, the end result is basic incompatibility that you are not aware of. We've talked at length about how humans, unlike the great tits or barnacle geese, or the rest of the animal kingdom, for that matter, have a fairly strong tendency to misrepresent ourselves to potential long-term mates. That is why, particularly in humans, an extended courtship (with lots and lots of that yapping that you see in the happy albatross couples) helps flush out such inconsistencies.

Behavioral compatibility is often a predictor of success. But what about opposites? Is there any situation in which it pays to be different? Yes, but the difference is one of complementarities, where partners still share common goals, but each partner specializes in a different contribution to the relationship and to achieving those goals. It is not detrimental for the relationship if one individual is better at a task than the other so long as both are trying to achieve an agreed-upon objective, and this requires similarity of purpose. What is one way to measure how compatible you are? If you are a cockatiel, it just might be how well you get along.

Cockatiels are outgoing and expressive parrots that many people have as pets. I used to have a phobia of birds . . . well, a fear of their beaks, to be precise. One little cockatiel, Ginger, cured me. Perhaps because cockatiels are such extroverted birds, Ginger has no boundaries. The first time I met Ginger she didn't give a hoot about my fear. She marched right up to me, hopped on my shoulder, and shoved her face, beak and all, into my face and began panting like a dog. I calmly asked her to give me some space, since we hardly knew each other. Ignoring my verbal request, she threw herself down into my lap, continuing to pant. My fellow humans informed me that Ginger really liked to be scratched under her chin. And so a love affair began.

In the wild, cockatiel love is a long-term affair. Cockatiels nest in cavities, and both partners coordinate incubation and feeding. They stay close to each other and copulate regularly, even when it's not the time of year for raising chicks. Couples that are more compatible are more affectionate with each other, spend more time together, have more sex, and fight less often.[16] This brings us to a crucial hurdle . . . *sex*.

SEX—EVERYTHING YOU WANTED TO KNOW ABOUT BONOBOS

Speaking of sex, we really can't talk about compatibility as mates without talking about sex. The importance of sex, and the quality of the sex, should not be underestimated. That is probably why there are so many books written about this topic. I hinted earlier that it is a myth that other animals only have sex to reproduce. And I just told you about cockatiels, which have sex all the time. Sex feels good, and in case you think it only feels good for humans, consider this: all female mammals have clitorises.

How can I state so unequivocally that animals engage in sexual behavior for pleasure? Masturbation, for one. Who knew cats masturbates? But my male cat definitely does. Seriously. Even as an animal behaviorist, I had no idea that cats masturbated, and I was surprised by his appetite. I had always owned dogs, and, well, we have all encountered (or have had the embarrassment of owning) that one dog that humps everything.

But cats? I described his behavior to my veterinarian when my cat was about a year old and he told me my cat was masturbating. He then assured me that many male cats engage in masturbation, which, incidentally, is quite entertaining to observe (discreetly, of course). My vet further remarked that many owners spray their beloved pets with water to discourage such behavior. I was horrified. I feel it is my duty to say this for masturbating cats everywhere: please stop doing this to your cat. Imagine if just before you were about to climax someone burst in, sprayed you with water, and shouted "No! Bad person!" This might bring you back to your teenage days and the fear of having your parents walk in on you while you were masturbating or having sex.

I can think of no better argument in favor of masturbation than the fact that it is ubiquitous. Cats and dogs do it. Horses, squirrels, birds, bears, elephants, bonobos, and chimpanzees all get in on the fun. Why? Because it feels good! Some animals use objects, if necessary. In the early years of the great ape sanctuary I worked at, it was temporarily located in an establishment that housed a menagerie of birds, including lovebirds. These lovely little parrots, featured on the cover of this book, incidentally, delight in masturbation. Bet that is not the first thing you thought of when you looked at the cover.

At this particular place, there were three lovebirds housed together, meaning one would always be left out of the fun. The third wheel did not simply accept its lot in life. Instead, he or she used whatever was available, whether branches, toys, ropes, or other items, to enrich its daily life. One day, as I was walking by, I overheard a woman complain to one of the staff. She was saying that something just had to be done about the offensive behavior of that bird. I remember thinking that perhaps if that woman masturbated more she might not be so worried about what a bird was doing. Granted, we don't want to be exposed to people masturbating in public, but if your sensibilities are offended by a wild (or domestic) animal doing so, may I suggest that you might be taking things a tad too personal.

In humans, we know that, aside from the pleasure, masturbation to the point of ejaculation serves a useful purpose for males. Sperm turnover. Human males store sperm for about two weeks and then excrete the sperm in their urine to make room for new sperm. While too much of a good thing can reduce overall sperm numbers, ejaculating regularly give a man healthier sperm (in terms of the amount of DNA damage). Sperm that hangs around for days gets old and accumulates damage. Masturbating clears out the old sperm faster and makes room for newer, healthier sperm.[17] Healthier sperm equals healthier babies. At least men have a good excuse.

There is no real information on whether human female masturbation is adaptive, even though it, too, is observed in other species. At the moment, the consensus appears to be that it feels awfully nice. A very scientific conclusion, don't you think? It is not clear whether other animals masturbate to the point of orgasm/ejaculation, though I have a sneaking suspicion that when my cat's eyes roll up into his head and his tail does that little shimmy he's pretty close.

Aside from masturbating, the second piece of evidence that animals have sex because it feels good is that they even have sex for the heck of it. It is a myth that animals only have sex for the purpose of reproduction. Aside from cockatiels, albatrosses have sex outside of the breeding season, presumably to strengthen the pair bond.

The list goes on. But bonobos, close cousins of chimpanzees, are probably the most well known for their sexual escapades. They have sex all the time and for every reason imaginable, not just to make babies. They do it in exchange for food, to resolve conflicts (make-up sex!), to practice, to comfort one another, and to strengthen social bonds. You name it and it is probably a reason for them to do it. Hence the nickname "Apes from Venus." Interestingly, among the great apes, bonobos experience the least amount of social strife. Perhaps if we had more sex there would be a lot less war, or at least happier couples?

Now that we have established that sex feels good across the animal kingdom, I should point out that being well matched when it comes to sex is just as important as other forms of compatibility for long-term mating. Another one of those hurdles. Dragonflies are one of my favorite insects. I am simply fascinated by them; I can watch them for hours—if I can keep my eyes on them. By the time you see the adult flying around, its life is almost over. Dragonflies are fast, flying upwards of thirty miles per hour. Their maneuverability is virtually unmatched, as they move sideways, backward, and hover in place.

Like peacocks, dragonflies play a prominent role in many cultures. Strangely, in American and European folklore, dragonflies are considered to be evil, sinister, and minions of the devil. When you look at Native American and Japanese cultures, however, the symbolism of the dragonfly is more positive, representing luck, transformation, love, and adaptability.

But what do dragonflies have to do with sexual compatibility? Being physically well matched is so important in dragonflies that male and female genitals must fit precisely, like a lock and key, or mating does not occur. This is like those new versions of speed dating, called "lock and key" parties, where the men have keys and the women have padlocks. What does this have to do with dragonflies? With so many different species of dragonflies flitting around doing spectacular air aerobics, there might be some confusion about who should be mating with whom. But, no, there is a way around this conundrum. A male dragonfly's genitals will only fit with the genitals of a female if she's from the same species. Indeed, this lock and key system is the way for a dragonfly to determine whether he or she is hooking up with the right partner.

In dragonflies, being well suited for sex may simply be about physical compatibility, but in humans there are many other sex-related hurdles to overcome. For example, let's think about duration. If you are someone who likes to take your time, not rush things, and you are paired with a speed demon, you could be very dissatisfied. This would be like pairing up a turtle with a lion. Turtles and tortoises have sex for hours at a time while lions don't last very long at all, yet they engage in regular quickies for three to four days straight, round the clock.

This is somewhat similar to those couples that don't leave the bedroom for days in the beginning of a relationship. A friend of mine went to Venice for her five-day honeymoon and basically had nothing to recount of the Venetian wonders except the joy of having a bidet in the hotel room! Obviously, from a biological standpoint, this behavior drastically increases the chances of impregnation of the female.

Just because you can do something for a long time does not mean you do it well. So how can one tell if things are going well? Orgasm. I am not a male, so I will not purport to know about how things feel on your end, gentlemen, but from what I do understand (and have confirmed with a male friend—thanks for sharing!), ejaculation and orgasm is not necessarily the same thing. I think there is a general perception, particularly among women, that male orgasm is not that complicated. You ejaculated, so you must have had an orgasm. It's like our sign that you were happy and satisfied with your experience.

But given that a man can orgasm without ejaculation, have multiple orgasms before ejaculation, or ejaculate without orgasm,[18] I propose that there is more to the male sexual experience than meets the eye. I suggested earlier that my horny little cat was having an orgasm, but how can I really know this? One of the wonderful things about human males is that they can self-report on their sexual experiences. I suspect that if I were to ask my cat whether he climaxed, he would look at me in that thoughtful way he has and wonder why I haven't fed him yet.

Is there any way for us to get at the answer? Like similar pleasure-reward cycles seen when it comes to food, neurobiology is starting to get a handle on what happens in animals, and on how this compares to humans. The cycle goes something like this: excitement or desire, engagement in the act, orgasm, followed by a recovery period. But since ejaculation occasionally does not equal orgasm in human males, how can we distinguish whether male animals have orgasms or simply ejaculate?

One clue is that the same reward center that lights up in the human brain when males have orgasms and ejaculate also lights up in the brains of rats when they ejaculate. In addition, if a male rat is sexually stimulated but not to the point of ejaculation, the "like" center of his brain shines. This "like" center provides the incentive to continue, presumably all the way through to activation of the reward center, or orgasm.[19] This is where the theory that male orgasm evolved because experiencing pleasure makes you want to keep doing something came from. Given that frequent mating helps ensure the survival of a male's DNA, scientists have long theorized that the orgasm evolved to encourage frequent copulation, since pleasure acts as a reward, conditioning organisms to continue to engage in the behavior with which the pleasure is paired.

What about females? I think we have gotten shortchanged on the orgasm front, and it seems that female orgasm has been shrouded in mystery or even treated as a myth. Worse, perhaps, is the notion that female orgasm is simply a by-product of the male orgasm.[20] I strongly disagree with this hypothesis. The fact that the same parts of the female rat brain get going during sex suggests to me that orgasms serve the same purpose in females: to encourage her to have more sex. Why do some scientists think that the orgasm is not important for females? Because women can get pregnant without having an orgasm. This is kind of insulting to women, and it introduces what I consider to be an unnecessary distinction. As I already mentioned, sometimes men also ejaculate without orgasm, which means that a man could get a woman pregnant without having an orgasm!

It is also a myth that the majority of women do not experience orgasms. The truth is that 90 percent of women have had an orgasm by the time they have reached the age of thirty-five, and 50 percent of women have at least one orgasm during any given sexual encounters.[21]

Females of other species also have orgasms. It is pretty evident in chimpanzees and Japanese macaques. Orgasms in macaque females may not be all that different from ours. By watching a male and female mate, researchers have noted that sometimes there is a dramatic change in the behavior of females. What does she do? She will turn her face to look back at the male while simultaneously reaching back and clutching on to him. She then experiences quivering muscle spasms while making specific types of sounds.[22] Hmmm.

In stump-tailed macaque monkeys one observes the same thing, but scientists went a step further and measured physiological responses of females during suspected orgasms. What happens to a macaque shows a similar pattern to what we see in human females: increased heart rate coupled with uterine contractions and spasms.

I am going to go out on a sexual limb here and state that the absence of an orgasm for some women during a sexual experience is not evidence that orgasms are not important for female reproduction. Given the lack of data to the contrary, I will rigorously defend the right of all females to have an orgasm, and I will reject the notion that the female orgasm is merely a byproduct of the fact that the clitoris comes from the same tissue that in males becomes the penis during prenatal development. If you read that last part again, you might very well be inclined to conclude that, but for our clitoris, which gives rise to the orgasm in females, males would lack the tissue that develops into a penis—the vehicle for male orgasm.

This then begs the question: when a female, who is clearly capable of having an orgasm, as most women are, does not have one, what does it mean? A possibility is stress, which inhibits and disrupts the sexual reward cycle in rats.[23] Stress probably has the same effect in people. Then there is simple distraction, or maybe overstimulation. The latter may sound counterintuitive, but it can happen to guys, too. And finally, the one reason that no one wants to talk about: poor-quality sex.

People rarely talk about sex. What do we do instead? We fake orgasms. I realize that when it comes to faking orgasms, the discussion usually focuses on women, but let's talk about fish for a moment. A female trout will select a spawning site to deposit her precious eggs. Just before she gets ready to release her eggs, she crouches, opens her mouth, quivers vigorously, and shouts, "Oh, yes!" Okay, she doesn't actually shout, but the rest it true. Since fertilization is external, the male shakes right alongside her.

In normal spawning she would then release her eggs, and he would release his sperm. The "fake" orgasm part comes in because sometimes, even after all this violent shaking, the female doesn't actually release her eggs.[24] The male doesn't realize this and releases his sperm anyway. Oh snap! Why does she do this? A few ideas have emerged. Maybe the female just wants it all to stop, not unlike in humans, as you will see below. Or perhaps the male misreads the female's cue and prematurely releases his sperm.

Men and women, on the other hand, fake orgasms for a variety of reasons. I'll tell you the one reason that few, if any, magazine articles list for why females fake it: we might be bored. Yes, bored. It's not just us. Bonobos can get bored, too. At least in bonobos if one partner loses interest, the other one notices and gives up and presumably does not feel bad about him- or herself.

Not so with some human males. They just press on, thinking they are doing their partner a favor, waiting for her to get there. Why might women become disinterested? Because they know it just ain't gonna happen. Another reason why women fake it is that they don't want to hurt anyone's feelings.[25]

And, by the way, continuously asking us if we have gotten there is not helpful to the cause. An interesting difference between men and women is that women will fake an orgasm so their partner will have his orgasm because women consider a male's orgasm to signal the end of sex.

Naturally, since we don't seem to talk to each other about sex, many women would rather pretend to have an orgasm than discuss why they aren't having one. But the blame does not fall entirely on women for failing to speak up. We live in a culture full of contradictions. We are inundated with sexual images, but openly discussing sexual experiences, wants, needs, and desires is discouraged—even with our own partners! On top of that, a lot of men think a woman's orgasm is a reflection of their performance, which, by the way, sometimes it is. Okay, it frequently is, but not in the way men might think. The problem with faking orgasms, in my opinion, is that doing so perpetuates the idea that it is simply "difficult" for women to achieve one, otherwise they wouldn't need to fake it. Perhaps if women would stop faking orgasm, we could all start talking about it.

I don't think we needed scientific experiments to find out what sexual positions favored female climax, but nonetheless, a few determined scientists discovered that just a handful of positions led to women achieving orgasm. They also discovered which positions better encouraged males to reach orgasm. Anyone want to take a stab at which ones favored male orgasm? All of them.[26]

If you happen to be a woman, don't despair—there's a trick to this whole orgasm thing. It is actually three-fold. First, a woman needs to know which positions will bring her to climax (this varies somewhat from woman to woman). The second step is telling your partner what you like and what stimulates you (for some reason this is often not easy for men *or* women). If you don't communicate this, your partner may just do what worked for other women, and these techniques may not work on you. Therefore, communication is a huge factor in the sex hurdle. The third part is the fun part. Both partners can try things out together until they've solved the mystery of the female orgasm together.

So, in the end, it is not so much that it is "difficult" (women can actually arrive at orgasm very quickly when the right things are happening), it is just that it takes a high level of communication between the two mates. Wow, talk about hurdles! If you manage to jump over this one, you will both be well on your way to the winner's circle.

Ladies, you are not off the hook here. The irony is that men fake orgasms for the same reason: they realize that reaching orgasm is unlikely or is taking too long. I don't know if men get bored, but like women, they can be dissatisfied, not in the mood, or not that into you, so they fake orgasms because they don't know how else to terminate intercourse without making you feel bad. Another reason? To cover for premature ejaculation or the fact that alcohol is preventing them from achieving orgasm.

For both men and women, the more disturbing reasons for faking orgasms include avoiding conflict or arguments, and feeling inadequate or worrying about creating feelings of inadequacy in one's partner. Knowing how to please your partner works both ways. Both men and women need to speak up, and their partners need to listen. Once again, communication saves the day—or the night.

As we saw in wandering albatross pairs, the members of which develop their own special language, just between the two of them, communication with a mating partner is critical to maintaining the pair bond. I cannot imagine that there would be any wandering albatrosses left if they didn't talk to each other, or if they told each other things that weren't true.

In addition to compatibility in personality, behavior, and values, sexual compatibility should not be ignored. Sexual satisfaction, or the lack thereof, can make or break a couple. For some species, if the male fails to properly satisfy the female, he's out. In cockatiels, individuals are more likely to solicit another individual if their partners are sexually unresponsive. The key is the three-point plan to sexual gratification for both partners. There should be lots of yapping and then moaning going on in long-term strategy bedrooms (or kitchens) everywhere. But this is not a book about how to have great sex. I'll leave that for you to happily figure out on your own.

8

THE THREE Cs: COMMUNICATION, COOPERATION, AND COMPROMISE

Let's assume that all has gone well and that you have found a mate. You are both in it for the long haul; you've both gathered your information; and you're both are ready to begin a committed, exclusive relationship. What now? What comes next is what I like to call the three Cs: communication, cooperation, and compromise. Relationships of all kind require the three Cs. From the birds to the bees, other species seem to understand this, and they have overcome this hurdle. Yet humans seem to still have a lot of trouble in these areas.

Just look around your local bookstore. You will see shelves full of books about communication: some giving advice about how to communicate better, and others explaining how critical communication is to relationships or the idea that much of our difficulty in communicating results from males and females *actually* speaking different languages. Whether or not that is true, and despite (or because of) the fact that we have a sophisticated language (by biological standards), our communication does often run amok. This made me wonder whether other animals have misunderstandings or manipulate each other in their communication. Is it only human relationships that have such problems?

If you need to cooperate and compromise, the inability to communicate will certainly get in the way of success. If you look at where a couple seems to have the most difficulties, they are usually the areas of money, sex, children, division of labor, and overall priorities. Not surprisingly all of these components (and certainly many more) require cooperation by both parties.

So is the foundation of a good relationship communication or common purpose? And if the cornerstone of a good relationship is the degree to which individuals can cooperate and coordinate their actions toward a common goal, why are we so often on opposing pages?

Do other species have this problem? Are geese couples waddling around arguing about which way to go, what to have for dinner, how to take care of the children, and precisely when they should migrate and how much stuff they need to bring with them?

This brings us to the art of compromise. We hear all the time that the ability to compromise is at the core of every successful relationship. By its very human definition, *compromise* means each party loses something. Is it the same for animals? And if not, how have they mastered the art of compromise?

Just like sex, these topics could comprise another book altogether. However, because they play such a prominent role in animal interactions, especially long-term partnerships, it is worth taking a peek and trying to answer some of these questions.

YOU JUST DON'T UNDERSTAND!

For the most part, communication in animals is designed to avoid conflict and promote cooperation toward a common goal. Have you ever watched two dogs play with each other? Dogs use very specific signals to convey the message "I want to play with you!" The most familiar one is the bow. Unlike behaviors such as bites, headshakes, and growling, which occur in other situations, the bow is used exclusively in play. Initially it is used to start a play session, but then it is also utilized during play.

During play, it is almost always used after one individual bites the other.[1] Normally, biting, along with head shaking, is a very aggressive signal. The bow is needed to remind the other, "Hey, I didn't mean anything by that bite. C'mon, we're still playing."

The function of this behavior is two-fold: (1) to prevent the play from escalating to aggression and (2) to signal the desire to continue playing. You could take your dog out right now with one of those play ropes and watch how he or she will predictably bow almost always after biting the rope, shaking his or her head, or even growling. This, of course, assumes your puppy is schooled in proper doggy etiquette.

Remember our apes from Venus, the bonobos? The ones whose lives are filled with sexual escapades? Not surprisingly they also play. They play alone, and they play with each other. Along with humans and many other primates, bonobos have specific facial expressions that send the message that their sometimes-rough play is all in good fun.[2] Bonobo play includes bites, summersaults, pirouettes, and tickling.

Speaking of tickling, have you ever been tickled just a bit too long? The usual outcome is that you end up shifting rapidly from laughter to anger. The bow in dogs and the relaxed smile used by bonobos are just two small examples of a world of animal communication that demonstrates that one is playing and not fighting. Animals, in fact, go to great lengths to avoid miscommunication. Why? Well, in these two examples, you can imagine the consequences if there were a misunderstanding. Things could get out of hand in an instant, and what had been all in good fun could become an all-out, physically dangerous fight.

Not unlike the playful banter common in human couples. If someone pushes things too far or hits too close to home with a joke, a humorous moment can turn into an unpleasant one with just a few misplaced words. If we assume for the moment that when a message doesn't get across properly it's an honest blunder, how do these mistakes come about? Are we just not as careful as animals when it comes to the art of communication? There may be many sources of unintentional error in communication, but I am going to focus on three of them: not listening, not giving enough information, and stress.

DID YOU HEAR WHAT I JUST SAID?

Let's start with not listening or paying attention.

Do you ever have the sense that your partner is not listening to you, which tempts you to speak gibberish or say outrageous things just to check? This topic came up in a conversation with one of my friends. Although Samantha loves her husband, he has this habit of "pretending" to listen to her. He even responds with "Uh-huh. Sounds good," leaving her clueless; that is, until he doesn't answer one of her questions because he wasn't actually listening.

Since I know her husband loves her very much, I can't help but wonder, what on earth is going on here? More than you probably realize, ladies; sometimes your husband isn't listening because he can't actually hear you. Ironically, this is especially true when your voice is stressed. I know, I know, you are standing right there speaking to him and he can't hear you? Right! But it's possible.

You know those whistles that dogs can hear, but we can't? Female voices are to men what those whistles are to people. See, the brain of the human male is wired to pay attention to lower vibrations, or frequencies. Female voices, because they are usually higher in frequency and intensity, particularly when upset, don't necessarily resonate the same for men as they do for other women. A bit like male green treefrogs that become desensitized to repeated intense sounds and pay more attention to lower vibrations.[3]

Men may have a propensity to habituate to repeated sounds

(you) that are higher than normal in intensity. On top of that, because men's brains process voices differently, when a man hears a voice, he has a tendency to compare it to his own voice, trying to get a picture of the speaker, probably trying to figure out if the voice is that of another male.[4]

In the case of Samantha, it isn't that her husband doesn't care, or that he wasn't trying to listen to her (okay, maybe not *all* the time), his failure to listen is mostly likely because men aren't attuned to women's voices in the same way that women are, and sometimes he simply doesn't hear her. This is not an excuse for all you men out there to ignore your mate. It may mean, however, that men need to develop signals with their partners to re-establish attention and communication. For women, perhaps a way around this is to experiment with altering the tone or pitch of your voice. Another, and arguable more entertaining, option? Play the call of a male green treefrog randomly while you are talking. Just an idea.

While differences in male and female vocal frequency can be one explanation for why one person "wasn't listening" or was "tuning you out," I have observed (and experienced) another more common reason. Couples talk over each other. My upstairs neighbors clearly have this problem. Their situation is more dramatic though, as they are usually shouting at each other, often simultaneously.

The part of the brain that produces speech is also responsible for processing language.[5] Even if you aren't actually talking at the same time but are, say, formulating the response in your head before your partner is done talking, you may as well be talking. Because we "hear" our own words in our head before we speak them, we aren't able to listen to what the other person is saying.

You may be convinced that you can multitask in this way, but I would simply point out that even most animal couples take turns talking. Like tropical boubous. Who can't adore an animal with a name like that? I confess that it makes me giggle just a little every time I read it or say it out loud.

Found in sub-Saharan Africa, male and female boubous look alike. Both are fairly similar in size and have white feathers on their rumps that aren't visible unless their tail feathers are in the up position. They get their name from the sound of their call: bou . . . bou or bobobobobo, as if they are saying, "I'm over here." Both the male and the female can initiate communication; it all depends on which song they are singing. However, once they start, males and females take turns with a precision that is impressive. Except for a few songs where overlap is a ritualized part of the motif (sort of like a chorus in a human duet), males and females don't talk over each other.[6] Hmmm. . . .

I can't really say whether miscommunication happens between individual animals as a consequence of not listening, but when, like people, animals have to deal with background noise, this leads to mistakes. This reminds me of a funeral I attended once. I haven't been to many funerals, and it was my first Roman Catholic one. Other than visiting the Duomo in Milan, it was actually my first time in a Roman Catholic church. Unbeknownst to me, there was a lot of standing up and sitting down throughout the long service.

At one point, the priest told the congregation to turn to one another and say, "Peace be with you." Suddenly, the congregation came alive, flooding the church with voices and handshaking. I, being unfamiliar with this tradition, as well as being seated far from the priest, concluded that the priest had said, "Pleased to meet you."

Despite thinking this was a bit odd for a funeral, I dutifully turned to every person around me, shook hands, and said, "Pleased to meet you, too." It was only after I'd turned to my friend, shrugged my shoulders, and uttered the phrase that I realized, by the look on her face, that my communication had gone awry. Needless to say, it took a tremendous amount of effort on her part to resist bursting into laughter as she explained what was really being said.

Though a bit embarrassing, this was a fairly innocuous error. Given the lengths to which animals go to avoid errors, living in close proximity to humans and our noisy tendencies affects more than just their ability to communicate properly. From mistaken identity to not being able to detect that there is a potential mate or predator around, noisy environments present a serious miscommunication problem for many animals.[7]

Another frequent type of inadvertent miscommunication is saying less than you think you have. A lot less. Just as with humans, conveying information is an important part of most animals' lives. One particularly important thing animals talk about is predators. Take prairie dogs, for example. When a prairie dog sends out the alarm that a predator is coming she or he packs a lot of information into the call. Prairie dogs say something like, "Hey! Watch out! Here comes Joe, that medium-sized, brownish coyote, over the ridge on the left, coming toward us at a steady pace" While they may not use these exact "words," they communicate pretty exacting information about the potential threat.

Unfortunately, we don't necessarily do the same thing when talking to our partners. A common complaint among couples is that at one time or another, one person expects the other one to read his or her mind. This may occur in even the simplest of encounters, such as asking a significant other to do something, say the dishes. "Honey, will you do the dishes?" To which the significant other may reply, "Yes, of course. No problem." But some time later the dishes have still not been done. When asked why, the response might be, "Well, I *will* do them. You didn't tell me you wanted me to do them right away." The one doing the asking may think this was obvious. Why else would one ask? The conversation may then turn into an argument about a lack of love (you should just *know*), appreciation, or helpfulness.

I can't help but point out that in this scenario, and countless others, the asker failed to actually say what they thought they had said. Not only that, but the asker goes so far as to think that what was meant was obvious. Prairie dogs don't just shout out "Hey! Watch out!" and assume that everyone knows it is Joe the coyote coming from over that ridge on the left. Yet this is exactly what we do. And the closer the pair, and the longer the relationship goes on, the more often we miscommunicate in this way!

When it comes to animal partnerships that last many years, strangely enough, we see the opposite pattern. Siamangs, sometimes called black-furred gibbons, are an old-world primate found in Malaysia, Thailand, and Sumatra. Famous for their loud, conspicuous songs that travel long distances through the forest, a bonded pair—two siamangs in a "relationship"—wake up and sing a duet every day. The song has three separate parts: the introduction, the organizing, and the great call. Consisting of barks, booms, and screams, the pair sings for about fifteen minutes, and they continue this ritual every year they are together. In fact, new couples need a bit of time before they get the whole song done right.[8] Once they do, however, they don't just bang out a few notes after a couple of years and call it a day. No, they go through the whole sequence—every time.

Lastly, sometimes, like in honeybees, there is an epic failure in communication that results simply from being stressed or tired. Now, you might be wondering, "How exactly do bees get stressed out or tired?" The answer is, when researchers keep them up all night by randomly rattling them awake.

I don't know if, like many of us, bees that stay up all night are actually cranky the next day, but they sure mess up their waggle dance. The waggle dance is how bees tell their beehive friends the location of food. Unlike the prairie chicken dance, you won't see this dance at the current late-night hot spots. Normally honeybees are extremely precise and specific when they give directions to other bees. Heck, I know a lot of people who can't give directions as precise even if they have gotten a full night's rest!

There are two parts to the dance: the waggle and the return. The waggle is a figure eight and then a turn to the right followed by another figure eight and then a turn to the left. If the bee moves vertically the message is that the source of food is directly toward the sun. How long the waggle part lasts signifies the distance.

When sleepy bees were allowed out to find food, they came back and dutifully performed their waggle dance. The problem was that tired bees were way off when it came to communicating which direction the food was.[9] They just don't communicate as well when they are tired or stressed. And the same is true for us. When we are sleep deprived a lot of things suffer, including how well we communicate ideas and understand what someone else is talking about. Since recent studies show that a huge percentage of adult Americans (and probably people all over the world) are not getting enough sleep and need sleep aides to fall asleep, there might just be something to this.[10]

LIGHTEN UP, I WAS ONLY JOKING!

Up until this point I have focused on communication difficulties that are largely unintentional. If we go back to the barbs and jabs that couples often trade with each other in jest, when someone goes too far they may say, "I was just joking." Maybe they were, or maybe they really meant what they said and are lying to cover their tracks.

People in relationships lie for all kinds of reasons, like Samantha, who told her husband she really did like the ring he bought her so she wouldn't hurt his feelings. Here I would like to zero in on one reason for lying common to animals and people: to manipulate others. If we think back to our rooster, he certainly did falsely announce he had food to manipulate the females into coming around.

For another example, let's go back to prairie dogs for a moment. I will never forget what happened one afternoon while I was watching prairie dogs at one of my favorite colonies. Sometimes things get rough out there in a prairie-dog town, and individuals fight over boundaries. On this particular day, two males, Antonio and Mr. T, were in a knock-down, drag-out fight. Now Antonio was my favorite, but he was losing this fight. There was a moment when, through the binoculars, I could see that this dawned on him as well.

What is a prairie dog to do when he is getting his butt kicked? Flee. However, Mr. T was right on his heels as Antonio tried to make his escape. And then Antonio let out an alarm call for a coyote, which stopped Mr. T in his tracks. Mr. T, together with other prairie dogs who heard the call, stood up on his hind legs (like prairie dogs do when a coyote call is uttered), looking for the coyote. Meanwhile Antonio kept running until he reached the safety of his burrow. Oh, and of course, there was no coyote.

Whether it is rivals that steal food or interrupt sexual activity, there is a lot to be gained by manipulating others through deception. What about individuals in long-term partnerships? Do lifelong albatross couples lie to each other? I don't think anyone knows for sure, but I can't see what they would lie to each other about. "No sweetheart, all those feathers *don't* make you look bulky!" seems like an unlikely conversation.

If they were to deceive each other about whether there was food or a predator coming it would hurt their chances of successfully raising their offspring. In other words, they don't have much to gain from deceiving one another. Then again, maybe the female Adélie penguins mentioned earlier do gush over the rock their main squeeze brings them, even if they aren't thrilled with it. Who knows?

Now, your first thought might be, *Well, do they cheat on each other?* After all, infidelity is a form of lying, right? True, and as we already mentioned, and will look at again in more depth in the next chapter, infidelity is common in even the most "monogamous" of species. But for now, regardless of whether it is the yammering of albatrosses or the duetting of tropical boubous, communication among animal couples is generally designed to avoid conflict and promote cooperation toward common goals.

Where there is miscommunication, innocent or otherwise, there could be conflict. As we know too well, if the expectations don't match or assumptions are violated, disagreements may follow.

In my experience, many human relationships start out with both people talking to each other endlessly. Yammering like those albatrosses, sometimes even developing their own language. So what happens? I am going to go out on my proverbial limb and say that I suspect the basis for most conflicts is not only poor communication, but starting out with, or ending up with, incompatible agendas.

LET'S WORK TOGETHER—"PADDLE ON THE LEFT, YOU IDIOT!"

If you can't work together on an objective that requires two to accomplish, you are not going to get much of anything done, or at least not as much as is necessary. Your nest is built, but it might end up with a massive hole in it. And if you happen to be going in different directions, you can't really call that working together at all. You've got a beautiful nest in one tree and your partner's just finishing hers in a tree down the block. They look good on paper, but the reality is, no family again this year.

After kayaking with a date, I knew unequivocally that we were not well matched. No additional information was required. Tandem activities, like canoeing and kayaking require the coordinated action of both parties. Most of our time was spent paddling in different directions, which essentially created the not-so-famous "tandem kayak death spin."

We probably should have taken lessons from the black-bellied wren couples, who, in order to efficiently defend their territory, synchronize protection of their space. How do they do this? As only black-bellied wrens can—by singing together! Unlike some of the unchanging duets that happen in courtship, when it comes to defending their territories these wrens are a bit more flexible. Sometimes the male really gets into his part, called a phrase, and goes on a bit longer. Regardless of who gets carried away, each has to be able to anticipate when the other will finish so that there is no gap between them but also no overlap. Interestingly, both the male and female will terminate the duet quickly if the coordination is not sufficient.[11] Now why didn't I think of that on my kayaking date?

Since I introduced you to the splendid fairywren earlier, it is only fair that I include the purple-crowned fairywren here. Found in Australia, the male has a striking purple patch on his head, bordered by black. Now, unlike other "monogamous" fairywrens that are notoriously unfaithful, purple-crowned fairywrens lives harmoniously in monogamous pairs characterized by a high degree of fidelity.

What sets these fairywrens apart? Unlike other species of fairywrens, where members of monogamous pairs spend their free time independently and in search of adventure with other fairywrens, this species devotes itself to cooperative defense of territory. Pairs form a cohesive unit, mostly spending time close together, and, once again, they coordinate their song.[12]

We can see a common theme developing here. In animal species that form long-term partnerships, there is a high degree of mutual cooperation (and often singing!). Until I started thinking about dating from this perspective, it would never have occurred to me to be on the lookout for how well I could accomplish a task, or cooperate, with a potential mate. Whether it's kayaking, giving directions while the other is driving, changing a tire, or even putting together the dreaded IKEA furniture, these are some of the simple situations where teamwork in couples can inexplicably and rapidly break down.

I will never forget the date on which I first realized how valuable joint tasks could be in terms of information gathering. It was my third date with this fellow, and I had just purchased a new kitchen table from IKEA, and I faced a conundrum. I couldn't get the box up the stairs by myself. My date arrived later that evening, and I asked if he would help me bring the box upstairs. Not only did he say yes, but he also carried it up all by himself. Then, much to my surprise, he suggested we stay in and put the furniture together. I think he might have even said, "This will be fun!" Huh?

After a split second, a little light bulb went off when I realized that this would be an excellent opportunity to assess how well we worked together. Unexpectedly, it *was* fun. We laughed and joked, and when I messed up he patiently explained what I was doing wrong, showing me how to do it correctly. I realized after this that Sharon and Randy, the couple that went off into the forest for three months, might have been on to something. Maybe that was a bit extreme, but the point is that it's probably a good idea to deliberately create situations where cohesive, coordinated action is required *before* one gets married.

Returning to the challenges of kayaking, legend has it that there is a river on Kauai referred to as the "river of divorce." Why? Kayaking down the river is a favorite activity of newly-weds, who clearly did not have the benefit of three months in the woods together before getting hitched, since they often end up screaming at each other, crying, or uttering the phrase "I hate you"—and meaning it—after having just said "I do" a few days earlier.

COOPERATION AND THE SLIPPERY TEST OF TIME

Maybe you are part of one of those couples that has done everything by the book. You've finally finished building your hole-less nest, and it is on to living happily ever after. No? Besides communication becoming more problematic for many over time, cooperation in couples can also seemingly erode. A common area of contention is household chores and division of labor.

In the beginning, things are blissful; no one says anything about the socks stuffed behind the couch cushions or the toothbrush upside down in the toothbrush cup, or, if they do, they're smiling and shrugging while saying it. Cooking and dishwashing are done touchy-feely together before racing for the bedroom. But then it happens. Well, different things can happen, but specifically *that* thing—children. Let's face it; kids don't do much to help around the house until they have, well, houses of their own. So the bulk of all the work that needs to be done to keep offspring growing toward maturity falls on the parents.

In general, and particularly in marriages not defined by

traditional gender roles, couples divide household labor more equally as the female increases her time working outside the home.[13] Not surprisingly, although gay and lesbian couples don't necessarily split chores evenly, they share chores more equally on average than do heterosexual couples.[14] As we all know though, perception is reality, and sometimes, regardless of the actual amount of work being done, one partner can feel as though he or she is doing the bulk of the work. Resentment can build and suddenly there is strife between two people rather than cooperation.

There are a few ways in which animals avoid this kind of trouble. One way is by adjusting their behavior according to the other's needs. It is thought that one of the reasons already-mated albatross pairs continue their elaborate display is to update each other on their status. When one comes back from foraging at sea after a long period of time, the display may function to tell the other what condition they are in physically. By doing this the one heading out to sea can adjust how long to be gone, coming back sooner if, say, his or her partner didn't find as much food as was needed.

In this way neither partner, nor their offspring, starves, and they continue to function as a cohesive unit, helping and supporting each other. Charlotte (remember the one who gets humped by dogs all the time?) has a pretty amazing husband. She works from home as a freelance writer and therefore usually takes care of more of the housework and cooking. Sometimes, multiple jobs come in simultaneously and her husband . . . well, he steps up to the plate and cleans house, walks dogs, does laundry, makes pasta with butter (his specialty), and gets the kids to bed.

At the beginning of their relationship things were not always so. She once told me that she had written a list of everything she did around the house for a week and then showed him the list. She wasn't mad about it (well, she did vent a little to me), but she did want to be appreciated for all the little things she did every day. A long time later, he did the same thing. She swears that just seeing what the other partner was doing—in writing—really works wonders.

To resolve their conflict, one that was based on what one or the other perceived as unfair allocation of chores and burden, Charlotte and Jacques used communication (written), flexibility, and score keeping. We've seen communication and flexibility in animals, but keeping score? Like most of what we've seen, humans don't have a monopoly on this one either. Animals also solve the problem of potential inequality by keeping score. Tit-for-tat, to be exact. Like it or not, keeping score is an effective way to make sure your partner is contributing equally to the interaction.

How does keeping score work in animals? Here are the rules: (1) never be the first to defect, where *defect* means failure to cooperate, (2) retaliate only after your partner has defected, where *retaliate* means do not cooperate, (3) carry out only a single act of retaliation then forgive (cooperate again), and (4) only use this strategy where you know you will interact *repeatedly* with a particular individual. Of course, if your partner defects (fails to cooperate) and never gives back after the initial retaliation-forgiveness rule, then you default to doing what your partner did on the previous move, and the partnership dissolves. All kinds of animals use this score-keeping strategy to achieve reciprocal altruism, an "I help you, you help me" form of cooperation that is highly resistant to cheating.

This is exactly what vampire bats do. Like snakes that sense body heat, the common vampire bat has heat-sensing receptors in its nose that help it find food; that is, blood. While many people find them scary, vampire bats know more about give and take than a lot of people.

Finding food is tough for vampire bats, and on any given outing, only 60 percent manage to find a meal. With blood hard to come by, a huge percentage are often in extreme danger of starving. To avoid mass starvation, the bats cooperate, adopting a life-saving strategy. A hungry bat can ask for and receive help. A fellow group member who returns to the roost after having successfully found food will regurgitate blood for its hungry pal. There is a catch though. If the one who donated food finds him or herself in need of assistance on another evening, the one who earlier received a blood donation has to reciprocate. If he or she doesn't reciprocate, his or her buddy finds a new partner—pronto![15] Cooperation in the form of controlled sharing keeps the whole colony alive. Of course, one might argue that a cheater could make his or her way through the whole colony before everyone figures out that he or she doesn't give back. But given that vampire bats need food every sixty hours, coupled with the fact that multiple bats contribute to feeding a hungry individual on any given night (and the bats recognize individuals), it wouldn't be long before everyone knew who *not* to help. No moochers allowed in vampire bat groups.

For the most part, animals have a strong sense of fair play when they are interacting repeatedly with the same individual. But how does, say a capuchin, figure out what is fair? Well, for Vulcan and Virgil, two capuchins, it was pretty straightforward. In an experiment designed to see if the pair would be fair, researchers gave Vulcan a rock while Virgil was given some nuts in a jar he couldn't open without a rock.

Separated by Plexiglas, save for a small opening, it didn't take long for Vulcan to pass the rock to Virgil. Vulcan then had to wait to see if Virgil could open the jar and share the nuts once he got to them. There were six nuts in the container. After Virgil accessed the nuts he did not take all of them for himself, nor did he take five, or even four. Despite the fact he could have decided that he worked harder to gain access to the nuts (after all, he did have to struggle with the rock to get to the nuts), he split the nuts evenly with Vulcan.[16]

Even more interesting, capuchins compare what they are getting relative to what another is getting for the same amount of effort, and they refuse to participate if they think something is unfair. Not only that, but if they see another individual get more for the same amount of effort, or worse, get more for less effort, capuchins will flat out refuse to do the task in subsequent rounds of testing.[17] They also don't work together again with someone who was unfair. Unlike some people who continue to put in effort and build resentment along the way, capuchin monkeys appear to have better boundaries.

Chimpanzees do the exact same thing, demanding equal pay for equal work. Just as an aside, I wonder what would happen in the United States, where the Workplace Gender Equality Agency reports that women on average earn eighty-four cents on the dollar compared to men, if all women stopped working instantly and simultaneously?

But back to relationships. You can see that we are wired to be very sensitive to how much effort we are putting into something, and we compare our effort to the amount of effort someone else is putting in. Like capuchins and chimpanzees (and Charlotte and Jacques), we generally dislike inequality. It is no surprise, then, that this is especially true in our romantic relationships. If giving in a relationship is like withdrawing from an energy bank, and one's partner is not giving back in kind, a pair will go into a deficit, and this just doesn't make biological sense.

And it's not just capuchins and chimpanzees that understand this, even fish are aware of what their partner is contributing and behave accordingly. Remember those stickleback fish, where the males wowed the females with their symmetrical spines? Well, when put in pairs and presented with a predator (safely behind a glass), individuals inspect the predator. If one of them consistently stays a little further behind, he or she is considered a defector and the other one adjusts its distance so as not to put itself at more risk.[18] Given our sensitivity to fairness, instead of asking why your partner is keeping score, maybe you could ask yourself whether you are contributing equally and cooperating.

I think we run into trouble for at least three reasons. First, humans seem to have a much harder time establishing what is fair, and as we all know, perception is reality, and if someone isn't feeling appreciated, he or she may start to feel things are unfair, even if that assessment is somewhat inaccurate. Regardless of whether both people work or one person stays home, it's probable that someone may perceive an inequality in the division of household labor. This could be because of different definitions of what constitutes "household" labor (for example, is childcare included) or because the difficulty level of different tasks varies. Then you just end up having two people argue about what is fair. Second, unlike the animal examples, some of us continue to give our full effort while simultaneously resenting the other person instead of immediately adjusting what we give according to what we receive. And third, I am not sure that we understand how to compromise the way other species do.

THE ART OF COMPROMISE

Here is a version of compromise where everyone wins. If you are a hermit crab, the rules are simple: (1) decide whether to compromise in the first place, and (2) if you decide to compromise, the best strategy is the honest one. An individual hermit crab may be in a house (shell) that is too big while another one may be in one that is too small. Empty shells are hard to come by, and everyone needs one, so what's a hermit crab to do? Tap another friendly hermit crab on the "shoulder" and see if he or she wants to trade. There can be no forced exchange, and no one offers an exchange that would take them further away from the size he or she authentically needs, so both crabs end the exchange with a shell that is equal to or better (for them) than what they had when they started.[19] What if more of us acted "crabby" and viewed compromise as a tool to achieve mutual gains rather focusing on the loss?

Another thing we hear all the time is that every relationship takes compromise—the ability to compromise is at the core of every successful relationship. The "art" part comes in being good at it, knowing when to do it, being fair, and not holding grudges after you've done it. Admittedly, this is not always easy. By its very human definition, *compromise* means that each party must first let go of something in order to gain an agreeable resolution. Literally. Merriam-Webster provides the following definition: "a way of reaching agreement in which each person or group gives up something that was wanted in order to end an argument or dispute." No wonder no one likes to do it! Of course, going back to the idea of the importance of common goals, the more agreement members of a couple have with regard to objectives, the less energy they have to devote to perfecting this precious art form.

In many cases, animals have no trouble compromising where there is no conflict of interest and the goal is similar for both parties. Navigation is one place where this plays out fairly frequently. When it comes to traveling together, animals, like people, have some decisions to make. Namely, when to leave and which way to go. Animals can reach a consensus about what to do and where to go in a few ways. In smaller groups, sometimes more experienced individuals lead the way, sometimes not. The key tends to be that the magnitude of the difference between opposing directions is fairly small.

Another way to compromise is to take a vote. Do animals vote? Yes! They do so by vocalizing, moving their bodies in a particular ways, and initiating movement. As groups becomes larger, the majority rules.[20]

But what about when you have a group of two? There is no majority possible in a group of two, so it all comes back to having a common interest. Animal couples want to prevent poor choices, find a good nest or food site, and successfully raise offspring. To that end, compromises have to be made. The reality is that there is very little research on how animal couples arrive at these decisions or settle disputes when and if they occur. I'll use this opportunity to suggest that it would be a very interesting area of research, and further exploration of this area could give us some insight into how humans can better accomplish the same.

Given the lack of data on compromise in pairs of animals, let's stick with the example of who's behind the wheel when it comes to which way to go. Animals recognize that giving all parties' preference equal weight does not always lead to the best outcome, especially if there is a difference in the quality of the information. Meaning, don't follow someone who may not know where they are going.

This reminds me of Samantha, who I mentioned earlier. A very experienced traveler, she acquires detailed information before making a decision about which way to go upon, say, exiting an airport. Her husband George, on the other hand, uses the wandering approach. When there are time constraints, the wandering approach is not as efficient, and since both parties have the same goal (to reach their next destination on time), the appropriate compromise would be to follow the "which way to go" strategy used by Samantha. If, however, the goal is to see who makes the decision, well, they might still be at the airport. Could this explain why some Canadian geese have stopped migrating?

Maybe part of the problem for us, when it comes to compromise, is that people are often asking for what they *want*, not what they need. Because we are sensitive to fairness, we may become less willing to compromise if we suspect that what is being asked for is not what is truly needed by the other party. In addition, as a colleague of mine pointed out, we have the added complication of being afraid of where we stand if we compromise, even if it's about something we care very little about. In such instances, it is easy to see how negotiations become less about the particular issue and more about perceived power. But power, and the abuse thereof, is characteristic of despotic relationships, not healthy relationships that are cooperative and egalitarian.

Despots are aggressive individuals who monopolize resources, prevent others from accessing those resources, and insist on making all the decisions. Despots are not the same as leaders. Rather, they are tyrants and dictators. Despotic decision-making systems usually emerge among animals (and people) living in unpredictable environments in which it pays for one or more dominant individuals to control access to all the resources.[21]

In nature we find both males and females acting as despots,

suppressing others. Rhesus macaque monkeys are notoriously despotic, living in groups characterized by extreme dominance hierarchies and intense competition. Dominant females harass and exert control over other females, and males do the same. When we look at many of the monogamous species that form long-term unions, these features simply do not appear to be present. As a result, I think we really need to ask ourselves why, if we are pursuing a long-term, monogamous partnership, do we engage in behaviors that are more consistent with a dictatorship?

So down with despots, and long live the three Cs! With lots of communication, cooperation, and compromise, you will be well on your way to happily ever after, unless. . . .

9

GETTING CUCKOLDED

Until my graduate training I never gave monogamy much thought. I never questioned the social convention. And when I was younger I believed the simple headline "Men Cheat Because They Are Biologically Programmed to Do So." After I decided to study mating systems I quickly discovered that this statement is misleading, and, as I have brought up throughout this book, it's only half the story. Nothing drove this home for me more than my date with Tom.

Tom worked in Internet technology, specializing in cybercrimes, and he was very enthusiastic about his work. We met for dinner, and before the appetizers arrived he began telling me how he caught his wife cheating on him with his best friend. He clearly was not past the betrayal, and he struggled to keep his composure as he relayed the story.

Mind you, I had not asked for details, and as he talked I recalled several other dates who had shared similar stories with me. I began to wonder why we are so invested in the belief that, by and large, it is men who cheat. If you ask any man out there (and believe me I have!), "Do you think women cheat just as much as men?" you will hear, "You bet they do! And what's worse, they are sneakier about it."

I also questioned my female friends about their commitment to monogamy. Several confided in me that they find monogamy more challenging than their partners do, and they predicted that they would be more likely than their partners to be unfaithful. So is there any truth to the equality of infidelity or do I just know some outlier females?

Given the secretive nature of cheating, it is very difficult to get precise estimates. The data that exists suggest that anywhere from 6 to 60 percent of people have been unfaithful at some point in their life.[1] That is a very wide spread, but one thing is consistent throughout the data: there is not a substantial difference between men and women.

For instance, in one study of heterosexual individuals in serious monogamous relationships, 23 percent of the men and 19 percent of the women self-reported that they had cheated on their partner.[2] Four percent is hardly a newsworthy difference. Gender may not matter, but the length and nature of the relationship does. The chances of both men and women cheating go up the longer a marriage lasts. In addition, if a couple is just dating or living together but not married, the rates of cheating are higher. But being married, it seems, only helps you in the short term. Satisfaction matters, too, but we'll talk more about that later.

Despite this research, we continue to be inundated with the cultural message and myth that *men* cheat because they can't help themselves. If that is true, but women do it as frequently as men, does that imply that women, too, are biologically predisposed to infidelity? What does this mean for monogamy? Is it biological? Is it cultural?

As I have done throughout this book, I propose we set aside morality and look to animals to see what's going on with "monogamous" couples. Perhaps if we examine unfaithfulness in animals by looking at when and why it occurs, we can get some clues about what's going on in humans.

TILL DEATH DO US PART

It has become quite fashionable for scientists to join the nonmonogamy bandwagon and go so far as to state that monogamy is not "natural." However, this is no less sensationalist than claiming that monogamy is the hard-and-fast rule. If it weren't natural, at least for some species, then we wouldn't find it in nature. If it were the rule, then there wouldn't be so much indiscretion.

As with many things, there is a lot of gray in this area. While it is true that relatively few purely monogamous species have been confirmed, they do exist. For one tiny mouse, it is monogamy or bust, and there is no bust. Up to eight inches long (most of that tail), the California mouse is a poster child for fidelity in mammals. When presented with sexually receptive potential partners, both males and females vigorously reject the extra mating opportunities. Study after study confirms that both males and females are faithful for life, which in this case is less than two years.[3] Now you might be thinking, *Well, no wonder they don't cheat. They don't live that long.*

Okay, fair enough. Then perhaps we should look to black vultures for a more realistic picture, since they mate for life and can live for up to twenty-five years. In Mayan hieroglyphics the black vulture usually appears linked to death, and it is occasionally depicted as attacking a dead person. The Mayan population most likely did not know that vultures happen to be one of the most monogamous creatures on the earth. Black vultures lay just two eggs each breeding season, with the male and female trading childcare duties in twenty-four hour shifts.

With so many hours away from home, do they get some action on the side? According to DNA testing, no. So what is keeping black vultures in check? There are eyes everywhere, ready to step in and keep everyone in line. Black vultures live in family groups and mate out in the open at the nest site. When relatives are around (and they usually are) they attack any individual that solicits sex from someone else's partner.[4] Can you imagine if we could only mate out in the open and in front of family members?

With the mouse, lab experiments combined with DNA testing confirm their loyal ways. But with other species, like black vultures, the evidence rests solely on DNA. I see one major flaw in characterizing any species as wholly monogamous based *exclusively* on paternity testing—it is not necessarily a reliable indicator of sexual fidelity. It presumes that mating happens only when the female is ready to reproduce, and we already know that many species, including humans, mate all the time. In other words, just because all of the offspring are related to a female's mate, it does not follow that he was the only male she mated with. DNA evidence, therefore, is merely a way to "catch" an individual animal in the act.

Human cheating rates vary tremendously, but, where it's possible to get data in an ethical manner, we don't see high percentages of nonpaternity by the primary partner. While nonpaternity varies from less than 1 percent to around 16 percent, on average, a man who thinks he is the father is *not* the father approximately 10 percent of the time.[5]

Given what we know about human behavior and nonpaternity rates, if I, as a scientist, were to describe the mating system of humans (at least in the Western world), I would classify it as socially, but not genetically, monogamous. This means that two people form a monogamous social unit, a pair, but, due to infidelity, the offspring may or may not be genetically related to the male partner.

YOUR CHEATING HEART

We can all understand the fascination with documenting that humans (and animals) cheat. It is a bit like the rubbernecking effect of a car accident. But even if we label humans as merely being socially monogamous, not every individual in the system cheats. This raises a very interesting question. What causes some individuals to be unfaithful and others to remain true? Birds were once heralded as the prime example of how "real" monogamy works.

We now know that, for the vast majority of birds that appear faithful, DNA testing has blown the whistle on them. Take swans, the iconic symbol of everlasting love. Who hasn't seen that photo of the two swans, beak to beak, necks curved into the shape of a heart?

A pair of swans can mate for many years, even for life. But next time you consider giving your mate a Valentine's card with that sweet heart-shaped photo, consider this: in any given clutch of eggs, 40 percent contain at least one offspring fathered by a different male swan. That means one in six cygnets, or baby swans, has a different father than it was "supposed" to have.[6] The reason? The current explanation is that female swans stray to ensure access to a fertile partner. I guess we should really be giving Valentine's cards with pictures of mice or maybe vultures.

Just as with swans, infertility can be a problem for human couples. If two individuals are closely related, the probability of infertility goes up. That is what was likely behind the low survival rate of children born to the kings of Spain during the Habsburg dynasty. As in many royal families, the Spanish Habsburg kings often married close relatives. King Charles II was the last king after successive generations of inbreeding. Nine out of eleven marriages were between individuals related at the level of third cousin or closer.

While over 70 percent of the children born of these marriages died before the age of ten, King Charles II survived, but he himself was impotent, mentally handicapped, and physically deformed. Despite marrying twice, first to Marie Louise Orléans, and later to Mariana of Neuburg, he produced no children.[7]

It is always risky for a woman, much less the wife of a king, to cheat (just look what happened to the wives of Henry VIII when they got caught), but if either of King Charles' wives had been a Lariang tarsier, the "bloodline" could have continued and no one would have been the wiser.

Often confused with the pygmy tarsier, this little nocturnal primate is found in Sulawesi, Indonesia. When I say tiny, I mean itty-bitty. Four inches long, with big saucer-shaped eyes, Lariang tarsiers weigh in at approximately a quarter of a pound.

In the majority of cases, they live in small groups that consist of a mom, a dad, and the couple's offspring. They sleep together and, you guessed it, they sing together daily. Usually the mom and dad are unrelated, and all the children are fathered by the dad. But when the two adults are closely related, female Lariang tarsiers look for love elsewhere.[8] Tarsiers are not the only species to avoid inbreeding by engaging extrapair mating. Considered socially monogamous, an Ethiopian wolf female will mate with other males in the neighborhood when she and her partner are closely related.

Granted, humans now have a global population to choose from, and it is unlikely that we are marrying our cousins, aunts, uncles, or siblings. So what else might influence how strictly monogamy is adhered to? What if it were simply a matter of where you live?

This may be something foxes, coyotes, and people have in common. Like most canine species, swift foxes were thought to be strictly monogamous. Also like many other canine species, they almost went extinct in the 1930s. Swift foxes are small, about the size of a domestic cat, and they live in prairies and deserts.

When it comes to the business of mating, things really get interesting. Sometimes they mate for life, sometimes they change partners every year, and sometimes there are three members to a group. A trio of swift foxes can be two males and one female or two females and one male. Regardless of the particular combination, in one population studied it was found that a female's mate did not father her pups 50 percent of the time.[9] For this species, the local environment has a strong effect on whether a population is socially monogamous or genetically monogamous.

Change the availability of resources and you often see a corresponding change in mating behaviors. When female animals have a harem of males it's called *polyandry*, as opposed to *polygamy*. Like the harem of females ruled by one silverback male gorilla, a similar picture is often painted of human societies, in which the practice of females having a harem of males is said to be virtually absent. But polyandry isn't just for those wattled jacanas and some swift foxes. It's for people, too.

I know, I know, the history books tell you all about the men who had harems, and there's not so much as a whisper about the reverse. While it *is* unusual (just as it is in animals), I say we settle this once and for all. A woman having multiple husbands is characteristic of over fifty human societies, a large number of them located in the Tibetan plateau.[10] This type of human mating system is found predominantly in egalitarian societies in which coalitions and team building are facilitated by leaders. What does this look like in practice?

A woman has exclusive sexual rights to more than one man, and all of the men contribute resources to raising all of the children born to that woman regardless of which man was the biological father. What is responsible for this reversal? It always seems to come down to who controls the goods.

As we have seen, males frequently control access to resources, and this sets the stage for the male mating behaviors we have discussed at length. But what happens when the environment doesn't support this approach? Maybe resources are very scarce and hard to secure. In such instances, it may be better for males to share a wife, especially if the woman inherits something valuable, like land.

It is thought that this explains polyandry in India, Sri Lanka, and Nepal, where land is hard to come by and large amounts of land are needed to support a family and maintain high social status. Similarly, if production on the land is low and men do all the labor, more than one man may be needed to provide the resources needed for a woman and her children to survive.

In other cultures, men behave a little bit like the lions of the Serengeti plains, where two or more men will form a coalition and share a woman. Why? If a man has to be gone for long periods of time, another man could steal his wife. The Inuit solve the problem by establishing arrangements whereby a man shares his wife with another man, occasionally his brother.

Both of these explanations defy what we are all led to believe: that the "traditional" polygamous male is forced by modern society to be monogamous, thus going against his very nature. Human nature contains it all, whether it is promiscuity, polyandry, polygamy, monogamy, or any combination thereof. Simple differences in the physical environment may be one reason for the patterns we see. This, of course, implies that human mating behavior is malleable, changeable, and not the fixed, rigid picture we paint in our history books.

Behavioral flexibility is often a sign that there is something going on at the genetic level. Is there a genetic component to monogamy? A mighty molecule called *vasopressin* is emerging as a player in the fidelity game, and prairie voles are showing us the way. Prairie voles, as I have explained, form lifelong monogamous partnerships, and they cooperatively and aggressively keep outsiders away, having little tolerance for interlopers, whether male or female.[11] A quick skip down to the meadow, and you get a whole other story. The meadow vole, a close relative, is the picture of promiscuity.

Why the difference? When prairie voles mate, vast quantities of vasopressin, a hormone, are released. Like the hormone oxytocin, vasopressin is present in all mammals, playing a role in bonding and social behavior. Interestingly, species-specific behaviors are linked to differences in the number and location of receptors for this hormone in the brain. This pattern may explain why different, but closely related, species of voles are monogamous or promiscuous. While scientists are still working out the details, it's known that vasopressin is released during mating and bonding into the brains of prairie voles at much higher levels than in meadow voles. Vasopressin floods the reward center of the brain, that place where orgasms are registered. If you block the vasopressin receptors in a prairie vole, it suddenly starts to look for a little vole side action, just like its promiscuous cousins.

Not surprisingly, a gene has already been identified that contributes to how many of these receptors an individual has. Evidence is emerging showing that the combination of genes a person has may affect the degree to which that person can form long-lasting social bonds, including pair bonds.[12]

Some factors, such as infertility, environment, and genes, may influence whether individuals mate outside their primary relationships, but there are behaviors that may also lead to infidelity.

HOW TO LOSE YOUR LOVER

If you don't pay enough attention, you could lose your mate altogether. Mate guarding, undertaken by males or females, is one way to ensure your partner remains faithful. If your partner is with you all the time, or you can account for his or her whereabouts and with whom he or she is spending time, you may have little to fear.

Years ago, before the age of cell phones, I knew a young woman who made her boyfriend call when he was leaving from anywhere and then call again as soon as he reached his destination. She claimed that she was worried about him driving on the dangerous Florida roads, but it was only too obvious to the rest of us that she was keeping serious tabs on her very good-looking boyfriend. I probably needn't add that he soon got sick of her hypercontrolling tactics and started dating her much more relaxed roommate.

Not everyone can live up to the high standards of a prairie vole, and not all mates can constantly guard each other. The reality is that in the majority of monogamous species, whether siamangs, cockatiels, albatrosses, or even humans, mates must spend some time apart. What mates do when they come back together (see each other at the end of the day, so to speak) can have a profound effect on their long-term success. Yearly anniversaries just don't cut it for French angelfish.

Favorites of divers and snorkelers because they are fearless, these fish make their home in the coral reefs of the western Atlantic Ocean. They mate for life, but spend only about 50 percent of their time with their mate.

Like other fishes, the French angelfish has external fertilization and spawns in the open water. During mating, the male and female swim close together, and the female releases thousands of eggs for her mate to fertilize. When not making baby angelfishes or looking for food, the happy couple swims side by side, cooperatively defending its rather large territory of approximately 5,200 square feet from nosy outsiders. That's about twice the size of the average American home.[13]

Since they don't spend every minute of the day together, after one or the other returns from being gone a long time, they circle around and around each other. Officially, this behavior is called "carouseling," and it presumably reinforces their bond.[14]

What do many of us do? When we come home from a long day's work do we twirl around each other? I suspect not, though maybe we should start. Warning, you should probably let your mate know ahead of time before trying this tactic out one evening. Or maybe not; it could be fun to surprise him or her. Instead of anything resembling fun, many of us begin by reciting a laundry list of things that didn't go right during the course of the day. "I had to wait an hour and a half just for a text answering my question about whether you were picking up Julian from preschool." This is in spite of the fact that research clearly shows that successful couples have as many as twenty daily positive interactions for every negative one.[15]

They may not dance, twirl, or sing, but couples that exchange compliments, affectionate touches, or knowing glances are more likely to make things last. If, instead, partners rush out in the morning, spend the day interacting with the world, come home exhausted, and gripe at one another, they are not likely to outlast even the "carouseling" angelfish couple, whose lifespan is roughly ten years.

Unlike angelfish that are together on a daily basis, some monogamous animals spend months away from each other. If you are an albatross or an emperor penguin, you take turns looking for food. On the other hand, if you are a mourning dove or a bald eagle, you and your partner may mate for life, but you only spend time together during the breeding season. The rest of the time, during migration, both males and females go it alone.

After the migratory period, both the male and the female navigate back to the nest, where they enthusiastically greet each other. Then, just like so many of the other animal couples we have discussed, while the pair is together for four to six months, the partners go through their repeated bonding rituals, complete with ecstatic displays.[16] Year in and year out, for as long as they are together, they will repeat these rituals over and over. Once again, there is no getting too comfortable, as they continually work to maintain and reinforce the bonds between them.

I think some human couples understand what seems intuitive for angelfish, bald eagles, and so many other species. To maintain a relationship, you've got to work at it. It doesn't always need to be fancy, but it does need to be consistent. Unlike bald eagles, you don't have to link hands and free fall, hoping that you can coordinate, at precisely the right moment, when to let go so you don't hit the ground.

Instead of free falling, you can keep it simple. A colleague of mine, Heather has what, on the surface, may seem like a boring set of rituals with her husband, James. Weekly, they have one night reserved as a movie night and another day when they spend half the day together with no one else. Their daily ritual, however, is centered around food.

Heather is an amazing cook, and both she and James are a bit like foodies. They work in the same building and often have lunch together—usually some fabulous creation prepared by Heather—and, while eating, talk endlessly about what meal they will have for dinner. They plan out every detail, including what time dinner will be ready. It's amazing to watch. While discussing their current meal and planning the next one, the partners are completely focused on one another. Their eyes light up and they are smiling, both looking forward to going home at the end of the day. To eat, I assume.

If food isn't your thing, not to worry, you have plenty of options. You could play cards, work in your garden together, or cuddle up and watch that television show you both like. Heck, clean the house or pay the bills naked if you want. Whatever you choose, as long you both enjoy doing them together, daily, weekly, and monthly rituals will continually reinforce your bond.

Sometimes life gets in the way. Kids happen. Bills and stress happen. Your partner calls for your attention, but you don't recognize the signs. That's bad enough, but if you fail to support your partner in a social interaction, it may just be the kiss of death. If we go back to our French angelfish, we remember that they defend their territory cooperatively, keeping both sexes out of the territory and out of reach. Interlopers are enemies of the couple, and the couple that stays together is the one in which neither the male nor the female side with the enemy.

For instance, have you ever complained to one of your friends about your partner when suddenly your friend exclaims, "What a jerk!"? Have you felt that flash of annoyance, followed by the desire to come to your partner's defense immediately? That's the common-enemy principle. Supporting each other, especially in aggressive interactions with outsiders, is item #1 on the Relationship 101 syllabus. French angelfish know this, and so do Bewick's swans.

I realize I may have fogged up the rose-colored glasses when it comes to swans, but for swans, even those unfaithful ones, coming to a mate's assistance is one predictor of success. In Bewick's swans, males and females that stick close together and come to each other's assistance in aggressive encounters with other swans fare better than pairs that don't.[17]

I am sure that I am not the only one who's ever been thrown under the bus by a romantic partner. I was dating this man, Peter, for about two months when the moment came, as it does in all new relationships, to meet his friends. We met two couples at a Middle Eastern restaurant and things appeared to be going pretty well . . . on the surface anyway. Halfway through the meal I asked one of the girls what she did for work, and she replied that she was a dancer. I was impressed and envious. After my dream of NFL stardom, my dream of being a famous hip-hop dancer was a close second.

I conveyed my enthusiasm and she began laughing, as did everyone around the table—including my boyfriend. At first I thought it was an inside joke, and then she said, "No, I'm a stripper," to which Peter then said, "You have to excuse my girlfriend, you know how gullible people with PhDs can be." Huh? An awkward silence came over the table, and even his friends looked embarrassed for him. He, on the other hand, not noticing, continued to happily scoop his hummus with a piece of pita bread.

Granted, the relationship was new and his friendships were

old; nevertheless, I felt the wheels of the bus go round and round, round and round. The damage was done and there was nothing left to do but to end the relationship.

I THINK I WANT A DIVORCE

In animals, if one's partner is siding with the enemy, say, an interloper, the relationship is already on very shaky ground. What could cause an animal to change its devotion to a mate? Do animals divorce? And if so, is it common?

If you are a cockatiel, you might find yourself longing for a different mate if your partner is not sexually responsive. Cockatiels are high on the monogamous scale, but they will stray if their partner fails to respond to them sexually. More often than not, if things aren't going well with a current mate, a cockatiel will start an extramarital affair and ultimately switch partners. Sometimes even animals make a mistake when choosing their first mate.

However, unlike the poor success rate of human second marriages that have their start in an affair, cockatiels usually do better the second time around. In the case of cockatiels, the original couple wasn't behaviorally compatible, thus the impetus for changing partners in an otherwise monogamous species.

There are a few other reasons, aside from overall compatibility, that explain why animals that form lifelong relationships get a divorce. Let me clarify that divorce in animals includes situations in which a living partner deserts another and/or there is active mate switching. It does not include animals that are widowed because of a mate's death.

The common tern is a seabird. Like the bald eagle, the common tern has a breeding season, after which partners go their separate ways, only to meet up again the following year. Same time, same place. In contrast to the cockatiels, who switch mates due to incompatibility, the majority of terns divorce because one or the other fails to show up on time.[18] The couples that remain intact do so because the companions arrive at the breeding grounds within two days of each other. But in pairs that get divorced, the mates arrive more than seven days apart from one another. For terns, too, it pays to be in sync with your mate.

Not all species in which members of a couple arrive at a breeding site at different times have high divorce rates. In the waved albatross the divorce rate is only about 7 percent. However, those individuals that arrive early, with their mate nowhere is site, are not above having a dalliance. Over 60 percent of individuals, both male and female, are unfaithful.[19] But why is the divorce rate so low? You might recall that courtship in albatrosses is an extended affair, lasting several years before a mate is chosen. Therefore, it is extremely costly to jump ship and look for another mate. Valuable years of reproduction could be lost. When divorce does happen, it is usually the result of a lack of reproductive success.

So, how does the human divorce rate compare to the divorce rates in animals? A survey conducted using data collected between 2006 and 2010 reveals that, in the United States, roughly 30 percent of marriages dissolve after ten years, and 50 percent after twenty years.[20] By comparison, common terns, for example, divorce approximately 19 percent of the time,[21] which is fairly high on the animal spectrum of separation rates. The average is somewhere around 10 percent, but divorce rates vary widely across species.

Just as in animals, the divorce rates in humans can vary widely across cultures. Among the Hadza hunter-gatherers in northern Tanzania, the rate of divorce for first marriages is roughly 80 percent.[22] A young Hadza woman will have little tolerance for infidelity, and like common terns, if her husband is gone too long, or there is gossip of an affair, she will simply replace him with another. Regardless of the reason, divorce is accepted. Similarly, divorce among the !Kung people is rarely contested.

Not so for Western society today, particularly if mate switching is involved. From allegations of violence to legal battles that cover everything from custody to property disputes, it is very costly to get a divorce in contemporary society. Personally, I have always found resistance to a partner's desire for divorce to be a bit strange. I suppose I see nothing to be gained from trying to retain a marriage with someone who no longer wishes to be in the relationship.

This made me wonder whether animals ever contest a divorce. It is almost criminal that, as a biologist, I have yet to visit the Galápagos Islands. It is on my bucket list, and when I get there I hope I have a chance to see the Nazca booby, a seabird. It's not as fancy looking as the blue-footed booby, but it's special nonetheless. Typically, Nazca boobies are monogamous, though sometimes there are grounds for divorce, and, in the majority of cases, it is the female who initiates and allows a takeover event by an intruding male, even assisting the enemy by being aggressive toward her former partner.[23] Even though her current partner vigorously objects to being ousted, ultimately he is forced to leave. Out with the old booby, in with the new.

TILL DEATH DO US PART REVISITED

By now you may have given up on the fairytale, so let me try to end on a positive note. In both humans and animals, a good portion of relationships do manage to stand the test of time, and some partners do stay together, remaining faithful for life. For such an individual, losing one's mate can be devastating.

So deep is the bond in geese that when a partner dies, the

one left behind cries out mournfully, puffing up its feathers and ceasing to eat. This grief can go on for months, and occasionally the mourning partner will die. Geese are so devoted that they will stand over the body of their fallen partner. I have seen this myself. Because of the grief I have witnessed by the mate left behind, I feel it is my personal duty to tell everyone to drive carefully if geese are walking across a road—they do not walk very fast. In case you needed a better reason to not run over a goose (or any other animal, for that matter), there you have it.

In the end, there may be more biological truth to the saying "people come into your life for a reason, a season, or a lifetime" than first meets the eye. As in animals, there is more variation in human mating systems and behavior than we often like to admit. Some people are diehard monogamists wanting only one mate for life. Others think monogamy with whomever they are with at the time is just fine. And still others are not so keen on monogamy at all. As with so many of the other factors involved in a successful partnership (for example, compatible personalities), like-mindedness and agreement on what shape your mating system should take is a major component in forming a successful partnership.

10

IN A NUTSHELL

When I started this project I believed that by using my well-developed animal-behaviorist skill set, I would unlock the clues to my own somewhat-muddled personal life. What I got was that and so much more.

One interesting observation I had was that many of us are looking for the same thing: to connect in a real and genuine way with another person. At least that is what we tell others and ourselves. Yet most humans are floundering like fish out of water. Many of our connections (romantic or otherwise) are shallow and confused. Since animals don't have all the hang-ups humans do, the wild seemed like a good place to start looking for clues. I had already done lots of research about what motivated animals in their search for mates, and I thought I could apply the same principals to people—using myself as the research subject. I have to admit it wasn't always easy keeping a professional distance from myself, but my scientific approach turned out to be well worth it.

Despite our complex nature, there are biological principles that shape our perceptions and experiences out there in the dating world, just as they do in the wild world. We are constantly evaluating the suitability and compatibility of potential partners. To do this successfully, we need to have good information. Paying attention to the accuracy of that information is critical to success.

A good part of the information we are processing, especially at the beginning of getting to know someone, is based on appearance. This is one of those places where we can't overlook the importance of biology in our romantic pursuits. Those physical features that we instantly find attractive (or not) are signals providing us important information about the reproductive and genetic potential of a prospective mate. And remember, those first impressions, made in less than a blink of an eye and based exclusively on physical or chemical aspects, are extremely powerful, and their power lasts a long time. Therefore, though you might have found someone who looks and smells perfect, keep in mind that the initial signals telling you if that person would make a good mate, in terms of genetic material, may not necessarily be (and often are not) reliable indicators of the *character* of the person in the long term. If then, you are looking for more than a quick fling, my advice is, after you find someone attractive, become just a tad Buddhist—at least for six months—so you can take a closer look beneath the surface.

Another observation I made along the way was that many people seemed genuinely uncomfortable with not only their appearance, but also themselves. Perhaps this comes from a fear of being rejected. Turning myself into a guinea pig had a truly unexpected side effect. Instead of focusing on and worrying about myself, I became an observer, a scientist. I had to get to know and accept myself as I am and then decide exactly what I wanted in a partner. Amazingly, once I let go of my emotional attachment to what someone might think of me, I was able to do this rather painlessly. It was a strangely freeing sensation. In other words, for my "experiment," I was not going to focus on trying to make sure the other person would like me. Instead, I would have to be 100 percent myself. I'd have to decide whether I liked the person. And I'd have to accept that one of us may be on a different page at the end of the date (or couple of dates). Of course, I knew that might mean being rejected.

Raise your hand if you've you ever had that date you thought went swimmingly, but the person never called you again? I don't know about you, but both my hands are raised. As it happens, I found myself on both ends of the not-calling deal. Why? Because it's hard and uncomfortable to reject someone directly. In the end, I decided it was better to go the route of Koko and Ndume and just be upfront about it. To boot, Koko, unaffected by being rejected, was able to have a deep friendship with Ndume. Why? I began thinking that it might just be a result of what I observed about how uncomfortable people are with what they look like or who they are. Was Koko able to accept rejection so easily because she did not feel badly about herself to begin with? Maybe, when we have a way to figure out exactly what animals are talking about with one another, we might be proven wrong, but for now it seems evident that animals do not suffer from low self-esteem or other neuroses (except those living in close proximity to humans or in zoos).

This brings us to another important point. Animals don't feel badly about who they are, and unless animals are hunting or being hunted by predators, they don't spend their lives trying to *look* like or be something they are not. Maybe we can take our cues from them. I mean, have you ever seen a squirrel trying to act like a swan to get the girl next door? Not likely. Squirrels spend their time perfecting everything about being a squirrel. So, for all those squirrels out there, if you find that you are more squirrel than swan, your best bet is to embrace those acorn-hoarding instincts in you and just be the very finest squirrel you can be. It doesn't make any biological sense to try to change yourself into a swan. You might be able to pull off the illusion for a few months, but as I learned from Ryan long ago, trying to be what others want is not a strategy for long-term success. You might be able to convince someone to date you and even start a relationship, but eventually the little acorn hoarder in you is going to come out, and then what do you do?

What this boils down to is that whether it is your appearance or your personality, you have to be comfortable with who you are, both inside and out. The good news for us humans is that the amount of variation that exists in our species, both in appearance and behavior, exceeds that of all other species. Yes, we are wired to have certain preferences, but even those can change over time. Take, for instance, how women are biologically predisposed to take size into account. Women today may still care about size, but the size of what? As we've seen, women who have the benefit of better economic conditions tend to choose less typically "masculine" men. For this biologist, and many other women, no longer is it the sizes of the brow, jaw, and biceps that count. Just as important is the size of a man's . . . heart! Surprise. That should put to rest those Napoleon complexes once and for all.

Okay, so suppose you are now well on your way. You are comfortable with who you are, have been honest with yourself, and know what you want. The next thing is to be honest with potential mates. Honesty is really the basis for every successful mating experience. Whether one is adhering to a short-term or a long-term mating strategy, lying about your intentions is damaging.

If a quick tumble in the hay is what you're after, then you will surely be able to find a mate who wants exactly the same thing. For those who are having trouble finding the right match, there are lots of dating sites out there, some for long-term relationships and others for no-strings-attached encounters, making it possible to find just the person for you.

With the advent of computers, the Internet, and dating sites, there is no denying that we live in a very different world than even just fifty years ago. We have evolved into a truly global species, one that has drastically changed and continues to change the habitat in which we live. With these changes have come modifications in our behaviors and our mate preferences. For example, the divide between gender lines is increasingly blurred; women, in cultures across the globe, are more and more often working out of the house, and men have become more involved in the raising of children. The biological result? Men have gotten choosier and women are choosing more "maternal" men. Marriage is becoming less inherently about children, and a wider variety of non-nuclear family options are becoming more acceptable.

What does this mean? Whether it is men being more feminine or women taking on more traditionally male roles, cultural values are evolving. With all the recent mixing and reconfiguring of traditional roles, I discovered that it really isn't easy being a woman or a man. If the blurring of gender roles is not enough to contend with, we also have to face the fact that the biological wiring of men and women *is* different. With everything that we seem to be up against, how can we possibly find a suitable mate, not to mention true love?

Okay, let's not get bogged down in the global domination of our species and its ramifications for our planet and ourselves. First things first. We've figured out that being honest with ourselves, deciding exactly what we want in a mate, and being honest with potential mates are the first steps. The key to the next step is finding a person with values and goals that are similar to your own. Whether it is scrub jays, barnacle geese, tropical boubous, or albatrosses, a pair consisting of individuals more similar to one another is usually more successful. The best part is that those traits vary; they are whatever traits are most important to you. For blue tits it is personality, but for barnacle geese it is body size. You get to make your own rules, but it goes back to knowing what your rules are.

Finding a compatible mate is just as important as avoiding a maladaptive one, The thing is, if you are trying be in a monogamous, cooperative relationship, violence and other destructive behaviors are not conducive to success. We don't see these behaviors in animals that form long-term monogamous pairings. From this biologist's point of view, this is a major clue that these behaviors are maladaptive. Of course, they are only dysfunctional if your goal is to be in a mutually cooperative, monogamous relationship. If, on the other hand, you're goal is to be a dominant, take-no-prisoners, despotic leader of a pair, then by all means narcissism is the way to go. Change your behavior, or change your goal. You cannot have it both ways.

One of the most interesting and disturbing realities I came across was the prevalence of violence in couples. Statistics indicate particularly high levels of violence in younger dating couples. For a time, I lived in an apartment below a young couple that embodied these statistics. Between the screaming matches and physically violent outbursts, perpetrated by both of them, I couldn't help but wonder, what on earth kept them together. Late at night, as I lay awake, privy to yet another one of their explosive fights, I often found myself wishing they would act like proper animals and go their separate ways.

Obviously my neighbors, like many of us, had started a relationship crippled by mutual incompatibility. The results were disastrous. That brings us to the next step. How do we go about finding a compatible partner? Once again, animals have the answer; we need only to look at barnacle geese and albatrosses. Dating. Remember that albatrosses, which can live well past the age of fifty, take a few years to test out potential mates. On the other hand, maybe it's a good idea to go about it like Sharon and Randy, and head off into the woods alone for a few months. If both people come out still speaking to one another, chances should be pretty good they will make it in the long run. Regardless of how you decide to go about it, mate sampling makes good sense if the goal is to establish a relationship that will stand the test of time.

For a relationship to make it over the long haul, it must be constructed on a foundation of good communication. Gibbons don't take communication for granted and forego their ritualized song. No, they sing it to completion every day. Maybe we could all do with a little singing to each other when we wake up in the morning! And when it comes to communicating with a partner, we should aspire to be like the boubous, who rarely "talk" over one another. If they did, how would they ever be able to listen to each other and know when it was their turn to speak? With all the stresses in life, communication can certainly go awry, but sometimes we, like honeybees, just need a good night's sleep (or a nap).

Cooperation seems downright hard for many people, but it is absolutely essential to the success of any long-term partnership. Whether it's paddling on the left or the right, coordinated action is more likely to be successful. As we saw with migrating geese, having a common, agreed-upon direction or goal in mind is of utmost importance in enabling you and your partner to get to your shared destination.

Sometimes though, compromise is necessary, but we must be sure to be fair. We saw that vampire bats don't take kindly to being taken advantage of; they will stop sharing when a partner doesn't reciprocate. On the other hand, Virgil and Vulcan showed us what can happen when we don't bicker about who did more—everyone gets their nut!

This brings us to our last important hurdle: monogamy. Naturally, not all lifelong partnerships are characterized by fidelity. Rather, monogamy is a moving target for many species, including humans. As much as humans love to label and categorize things, putting ourselves in a box called *monogamous* is, as far as I am concerned, unproductive. There exists a tremendous amount of variation in this experience of being human. Looking around, it is obvious that individuals vary in height, weight, skin color, and sexual orientation.

The degree to which an individual is attached to, or participates in, fidelity has a connection to biology. When we look around the world, there exists a range of mating systems in humans, from strict monogamy to polyamory. Differences in human mating patterns and behavior cut across genders, and these differences have changed over time. Similarly, we see this variability in other animals traditionally classified as "monogamous."

If you were honest with your mate in the early stages of dating, making clear that monogamy is essential to you, then it would seem to follow that if you find that your current mate is not upholding his or her end of the agreement, you have the right to end that partnership. The same would be true for any trait that is deemed critical to mate compatibility.

Animals divorce. People divorce. It happens. Despite the extensive time commitment they invest in finding just the right partner, albatrosses realize that they have made the wrong choice, and they divorce. They, along with other monogamous species that form long-term relationships, understand that staying with a poor-quality mate takes precious time away from pursuing a more harmonious and fruitful union. If things aren't going well for a cockatiel couple, the partners simply switch mates. More importantly, both parties do not have to agree to separate; it's enough if just one of them wants to leave.

Divorce can be costly, for humans and animals alike, but it's not equally detrimental for all species. If you are an albatross, it may be years before you find another mate. If you are a cockatiel, you have likely already identified your next mate before you leave your existing partner. Some of us are like albatrosses; some of us more like cockatiels.

Perhaps the most profound realization I had was that everything I was reading, seeing, and experiencing during this journey challenged the messages I received when I was growing up. In the end, I was forced to throw out many of the myths that I had accepted as reality. The Cinderella myth, which really applies to both sexes, was responsible for much of the heartache I experienced.

Scientific studies in the past decades have unearthed a whole host of revelations that challenge so many of the social norms that we believe without question. Just the fact that we are indeed biologically wired to judge people by what they look like is, in itself, eye opening. We've also seen that women are equals to their male counterparts in terms of initiating physical violence (not sexual), especially within young dating couples, and we have also encountered the idea that boys and men may just be more emotionally fragile than girls and women. We have looked at the statistics that show that women (when not mate guarded) cheat just as much as their male counterparts, and we have talked about how women and men have similar desires to pleasure and be pleasured—not to mention very similar anatomy to do just that.

We've even seen that self-pleasuring is common in the wild kingdom. So, while new studies are showing that male and female brains work differently, the end result seems to be pointing to the reality that women and men aren't so different after all. And even more revealing is the fact that humans are not so different from our animal cousins.

Researching and writing this book has been a wild ride. I learned an enormous amount about myself, and all the men, women, and animals that surround me. But even more important, my in-depth look at the world around me strengthened my belief that it is our duty to question social norms, cultural values, and the all-too-readily accepted myths about men and women.

My biggest realization is that we are all interconnected through biology. Me to you to him to them, and it is probably time we embrace our wild connections and feel part of our whole living planet. Maybe this is actually the wild connection we are all looking for. That leaves us with only one remaining question to answer: Did I find Mr. Right? Well, I'm afraid I am going to leave you wondering. . . .

ACKNOWLEDGMENTS

An enormous number of people have contributed to making this book a reality, and I am certain I will fail to mention many of them. First, thanks to Uwe Stender for believing in me. Along the way several individuals read various incarnations of the early chapters, listened to me relentlessly talk about ideas in bits and pieces, and helped me bring the vision into focus. For specific comments and discussions, a special thanks goes out to Gisele Kirtley, Harvey Whitehouse, Martin Burd, Rob Dunn, and Jeremy Van Cleve. For reviewing the manuscript, checking for errors, and offering helpful comments and suggestions, I am grateful to Aditi Pai, Ramona Walls, and Erika Tennant. Thanks also to everyone at Prometheus Books who worked to make this book a reality.

Naturally, I thank my mentors Drs. Con Slobodchikoff and Charlie Janson. I think I may have been the luckiest graduate student in the world to have had two amazing people guide me toward achieving my goals. You were very different advisors, yet you both let me blaze my own trail. Thanks for having faith in me. I hope to always make you proud.

Thanks to everyone at NESCent, the National Evolutionary Synthesis Center, for providing me a home while I completed this book. I am grateful to continue to be a part of the NESCent family, surrounded by amazing colleagues and mentors. Allen Rodrigo, Susan Allen, and Craig McClain—I have been lucky to be mentored by you. Joseph Graves, as a mentor and a friend, your constant support and encouragement carried me through many challenging moments. Candace Brown, thanks for your friendship, support, hilarious visual ideas, and tolerating me popping into your office unannounced, spouting off about some crazy idea I had.

For all the coffee and interesting conversations, and for answering some of my very "inappropriate" questions, thanks to all the baristas at my local Starbucks: Justin, Drew, Grace, Emily, Enid, Nadia, Shasta, Burke, Henry, Jane, Taryn, Kate, Kate B., Caleb, Chris, Jared, Will, and Savane.

I have been lucky to have many friends rooting for me, providing me with a constant stream of encouragement, and discussing with me various thoughts and ideas, especially when I was frozen with writer's block. Thanks to Glenda Kennedy, Elizabeth Sbrocco, Mary Barrera, Ron Anders, Liliana Dávalos, Jesse Bering, Stacey Biasetti Lunde, Greg Chambers, Stacey Tecot, Melissa Mark, Katya Stubing, Julie Meachen, Ricky Burch, Frank Marí, Caroline Lin, Cookie Marano Biasetti, Maryam Mohaghegh, Robert Ziems, Augustine Romano, Jessica Hagan, Craig Bruce, and Diane Charland James.

Thanks to mom, my brother Michael, and my sister-in-law Laarni for your support and encouragement. And to my amazing niece and nephew, Mia and Mikey: you are two of the coolest kids I know.

Lisa and Dan Colvard, between the talks over lunch, brunch, e-mail, and text, combined with the awesome illustrations, your support and friendship have been invaluable.

Thank you, Alison Hill, for your friendship and for sharing your expertise, not to mention for telling me repeatedly and with complete confidence that I could, in fact, write this book. You are a smart, fantastic woman and a great friend.

A special thank-you to Patti Ragan and all the apes at the Center for Great Apes, with a special pant hoot mention to Kenya, Noelle, Christopher, Pongo, and Knuckles. Patti, you gave me my first opportunity to be in the presence of many magnificent individuals and to be a part of realizing your dream of providing a safe haven for so many. I am forever grateful for the gifts you have brought to my life.

Jason Sager, without you, this book literally would not have happened. Not only did your thoughts, comments, and hours of discussion with me vastly improve the quality of the manuscript, but physically I would not have been able to complete this project without your help. Carolyn S. and Michelle C., you are right alongside Jason in having helped me get through this process. Thanks for being the wonderful healers that you are.

And finally, to my person (you know who you are), thanks for making this book better than it ever could have been without you. You helped me beyond measure; you kept me sane; you made me laugh, pushed me through barriers, and held my hand all the way.

NOTES

CHAPTER 1. THE BIRDS AND THE BEES

1. For details on New Mexican whiptail asexual reproduction and "fake" sex see Manning et al. (2005).

2. To learn more about the account of hammerhead asexual reproduction see Chapman et al. (2007).

3. For more information on the net human population growth rate see Worldometers, http://www.worldometers.info/ (accessed February 10, 2014).

4. For more details on praying mantis mating behavior see Lawrence (1992) and Gemeno and Claramunt (2006).

CHAPTER 2. FIRST IMPRESSIONS

1. For information on how female blue tits adjust feeding effort see Limbourg et al. (2004).

2. For the time it takes to assess the appeal of a website see Lindgaard et al. (2006).

3. For information on evaluating faces see Willis and Todorov (2006) and Schiller et al. (2009).

4. For details on the persistence of first impressions see Sunnafrank and Ramirez (2004).

5. For details on emperor penguins see Groscolas (1986).

6. For information about body condition, preferences, and feeding effort in red-legged partridges see Alonso-Alvarez et al. (2012) and Mougeot et al. (2009).

7. For information on the limbal ring see Peshek et al. (2011).

8. For general references on plumage coloration in birds and foraging ability see Garcia-Navas et al. (2012); for reproductive condition in blue tits see Doutrelant et al (2008) and del Cerro et al. (2010).

9. For more information on peacocks see Gadagkar (2003), Hale et al. (2009), Loyau et al. (2007, 2008), and Petrie et al. (2009).

10. For details on mating rituals and symmetry in medflies see Hunt et al. (2002, 2004).

11. For details on reproductive success and symmetry in sticklebacks see Mazzi et al. (2003).

12. For facial symmetry perceptions in macaques see Beck et al. (2005) and Waitt and Little (2006).

13. For facial symmetry perceptions in humans and primates see Boulton and Ross (2013).

14. For variations in perception during the menstrual cycle see Little and Jones (2006).

15. For details on face shape and reproductive success see Pflüger et al. (2012).

16. For details on symmetry, attractiveness, and infidelity see Wade (2010).

17. For preferences for symmetrical faces by children see Vingilis-Jaremko (2013).

18. For the relationship between symmetry and health see Little et al. (2012), Scheib et al. (1999), Sefcek and King (2007), Thornhill and Gangestad (1999), Thornhill and Grammer (1999), and Van Dongen et al. (2009).

19. For references on asymmetry and semen quality in people see Firman et al. (2003).

20. For symmetry and semen quality in antelopes see Roldan et al. (1998) and Gomendio et al. (2000); for insects see Farmer and Barnard (2000).

21. For costs of being asymmetrical in the wood mouse see Galeotti et al. (2005).

22. For information on chin shape and attractiveness see Thayer (2013).

23. For cross-cultural preferences in face shape and attractiveness see Daniel et al. (2012). For information on fertility and feminine features see Pflüger et al. (2012). For human mate preferences see Pisanski and Feinberg (2013).

24. For the role of teeth as a signal of quality see Hendrie and Brewer (2012).

25. For details on what men find attractive see Dixson et al. (2010a, 2010b, 2011a, 2011b), and Singh et al. (2010).

26. For the link between male pattern baldness and heart disease see Yamada et al. (2013).

27. For baldness and perception of virility see Gosselin (1984). For baldness and androgens see Hamilton (1942).

28. For hair coloration in lions see West and Packer (2002) and Hill and McGraw (2012).

29. For hair coloration in humans see Hinsz et al. (2013).

30. For more on facial hair and women's perception of attractiveness see Dixson and Vasey (2012), Dixson and Brooks (2013), Dixson, Tam, and Awasthy (2013), and Neave and Shields (2008).

31. For details on the history of beards see Peterkin (2002) and Whithey (2013).

32. For waist to hip ratio see Wade (2010).

33. For information on evolutionary perspective on the function of physical attractiveness see Berry (2000), Hume and Montgomerie (2001), and Little et al. (2008, 2011).

34. For high heels see Morris et al. (2013).

35. For more on sexual swellings in primates see Alberts and Fitzpatrick (2012).

36. For details on brain mapping and position of feet relative to genitals see Ramachandran and Blakeslee (1999).

37. For mate recognition in lemmings see Huck et al. (1984).

38. For more on the role of odor in attraction and individual recognition see Brennan and Kendrick (2006), Roberts et al. (2011), Roney and Simmons (2012), Thomas (2011), and Milinski et al. (2013).

39. For MHC in humans see Wedekind and Penn (2000) and Lie et al. (2010).

40. For MHC-dependent mate choice in mice see Penn and Potts (1999).

41. For details on the role of major-histocompatibility-complex genes and mate choice in primates see Huchard et al. (2013). For a comprehensive review of MHC and mate choice see Havlicek and Roberts (2009).

42. For MHC and orgasms see Garver-Apgar et al. (2006).

43. For MHC and infidelity see Garver-Apgar et al. (2006).

44. For MHC and fertility see Beydoun and Saftlas (2004).

45. For more on how the brain processes smells (pheromones) see Lanuza et al. (2008), Moncho-Bogani et al. (2005); in humans see Belluscio et al. (1999).

46. For birth control and MHC see Roberts et al. (2008) and Havlicek and Roberts (2009).

47. For details on the study of intelligence and mate self-esteem see Ratliff and Oishi (2013).

CHAPTER 3. FALSE ADVERTISING

1. For details on strategic deception in baboons see Silk et al. (1999).

2. To view the report by the Beauty Company see TBC facts figures and trends at http://www.thebeautycompany.co/downloads/Beyer_BeautyNumbers.pdf (accessed February 14, 2014).

3. For information on rooster false advertising and comb-size preference see Marler et al. (1986) and Navara et al. (2012).

4. For false vocal advertising of green frogs and female preference for territory size see Bee et al. (2000).

5. For information on the history of cosmetics see Marceau (1985).

6. For details on reasons for hair removal see Hansen (2007).

7. For more on facial hair and women's perception of attractiveness see Dixson and Vasey (2012), Dixson and Brooks (2013), Dixson, Tam, and Awasthy (2013), and Neave and Shields (2008).

8. For more on the naked mole rat and parasites see Pagel and Bodmer (2003).

9. For details on when removing underarm hair became popular in the United States see Hansen (2007) and Lee and Ladizinski (2013).

10. For information on historical views of pubic hair see Hansen (2007).

11. For more on antler production and fighting see Currey et al. (2009) and Moen and Pastor (1998).

12. For details on fiddler crab deception see Backwell et al. (2000).

13. For information on deception in orchids see Schiestl and Cozzolino (2008).

14. For more on firefly deception see Lloyd (1984).

15. For fake tails and reproductive success in widowbirds see Andersson (1982) and Pryke and Andersson (2005).

16. For information on experiments with collared flycatchers see Qvarnström et al. (2004).

17. For details on colored rings and zebra finches see Pariser et al. (2010).

18. For information on the color red in rhesus monkeys see Khan et al. (2011); for red in humans see Hill and Barton (2005).

19. For details on studies with fake feathers and zebra finches see Witt and Caspers (2006).

20. For more information on female preferences in guppies see Hampton et al. (2009).

21. For information on breast size, fertility, and symmetry see Moller et al. (1995) and Singh et al. (1995).

22. For more on flower color, mate choice, and female preference in splendid fairywrens see Mulder (1997) and Dunn and Cockburn (1999).

23. For information on female preference and bowerbird construction in satin bowerbirds see Doucet and Montgomerie (2003).

CHAPTER 4. SORRY GUYS—SIZE MATTERS

1. For information on fighting success in longwing butterflies see Peixoto and Benson (2012).

2. For details on the account of combat in vipers see Shine et al. (1981).

3. See Kaufmann (1975) for more information on eastern grey kangaroo fighting rituals.

4. For information on how female presence influences leaf-cactus-bug fighting see Procter et al. (2012).

5. For information on how female presence influences treefrog fighting see Reichert and Gerhardt (2011).

6. For more on the winner effect in humans see Fuxjager et al. (2009) and Lopez and Fuxjager (2012).

7. For details on the winner effect in crickets see Reaney et al. (2011).

8. For specifics on brown trout size and mating success see Jacob et al. (2007).

9. For more information about male crickets and protection see Rodríguez-Muñoz et al. (2011).

10. For information on harlequin ducks see Squires et al. (2007).

11. For more on the effect of removal of male for birds see Fedy and Martin (2009).

12. For details on bottlenose dolphin kidnapping see Connor et al. (1992).

13. For an account of the story of Mick and Julie see Pines and Swedell (2011).

14. For more on infanticide in Grévy's zebra see Sundaresan et al. (2007).

15. See Breuer et al. (2012) for details on silverback size and reproductive success.

16. For female preferences on height and body size/shape in humans see Dixson et al. (2010a), Evans et al. (2006), Swami et al. (2008), and Fan et al. (2005), Courtiol et al. (2010).

17. For more information on turtle penises see de Solla et al. (2001).

18. For information on acorn barnacle penises and reproductive behavior see Hoch (2010).

19. See McCracken (2000) for details on the Argentinian duck.

20. See Hotzy and Arnqvist (2009) for specifics on the structure of seed weevil penises.

21. For information on human penis size, shape, and function see Gallup et al. (2003, 2006) and Bowman (2010).

22. For details on human male behavior after separation from a partner see Shackelford and Goetz (2007).

23. For more information on sperm structure in the European wood mouse see Gomendio, Cassinello, and Roldan (1991).

24. See Diamond (1986) for a discussion on testes size and sperm production in humans.

25. For more details on aggression in desert gobies see Svensson, Lehtonen, and Wong (2012).

26. See Gundarson (1965) for information on performance and self-esteem of short men.

27. For information on Napoleon see Asprey (2001).

28. For details on Japanese macaque orgasm rate see Troisi and Carosi (1998).

29. For more information on cowbird intelligence and reproductive success see Gersick et al (2012).

30. See Keagy et al. (2011) for details on bowerbird problem-solving abilities and reproductive success.

31. For more information on human preferences see Prokosch et al. (2009) and Stanik and Ellsworth (2010).

32. For more on friendship and reproductive success in macaques see Massen et al. (2012).

CHAPTER 5. ON BEING A CHOOSY FEMALE

1. For specific information on prairie chickens see Robel (1966).

2. For details on human male dancing and female preferences see Hugill et al. (2009, 2010).

3. For more information on pied flycatchers and nest site preferences see Alatalo et al. (1986).

4. On the trading of meat for sex in chimpanzees see Gomes and Boesch (2009).

5. See Yoda and Ropert-Coudert (2002) and Hunter and Davis (1998) for details on Adélie penguin preferences for rare stones.

6. For the amount spent on rings see Cronk and Dunham (2007)

7. For details on trading tokens for sex in capuchin monkeys see Chen et al. (2006).

8. For information on female preference for male fish with nests see Jamieson (1995) and Kraak and Weissing (1996).

9. See Waynforth (2007) and Place et al. (2010) for research on the attractiveness of married men and mate copying.

10. For specifics on kidnapping in bonnet macaques see Silk et al. (1979).

11. For more information on female competition in tree swallows see Leffelaar and Robertson (1985).

12. See Woodroffe and Macdonald (1995) for details on reproductive suppression in badgers.

13. For details on role reversals in wattled jacanas see Emlen and Wrege (2004).

14. For the evolutionary perspective on human female physical aggression see Burbank (1987).

15. For more information on the women of Kashmiri see Burbank (1987).

16. See Burbank (1987) for a discussion of verbal arguments in Ona women.

17. See Mehl et al. (2007) for details on the number of words spoken by men and women.

18. For details on domestic violence rates by gender see Straus (2004).

19. For more information on adzuki beetle mating strategies see Harano et al. (2012).

20. See Pitcher et al. (2003) for guppy experiments.

21. For mate sampling in humans see Brand et al. (2007).

22. For information on poison dart frog mating behavior see Meuche et al. (2013).

23. For more information on multiple mating in tropical water pythons see Madsen et al. (2005).

24. See Magnusson (1979) for details on the Javan wart snake.

25. For details on the longevity of human sperm see Birkhead and Moller (1993).

26. See Greiling and Buss (2000) for information on the rate of orgasms in women having affairs.

27. Additional information on langur infanticide can be found in Borries et al. (1999) and Wolff and Macdonald (2004).

28. Details on the health consequences of children born to older men can be found in Heffner (2004), Thacker (2004), and Le Page (2014).

29. For more information on when human males and females hit their sexual peak see Barr et al. (2002).

30. For specific details on fertility, miscarriage, and genetic abnormality rates in human females see Heffner (2004).

31. For additional information on menopause in animals see Sherman (1998) and Walker and Herndon (2008).

32. For details on the stalking, rates of violence committed by women

versus men, and rates of injury, see Archer (2000), Purcell et al. (2001), Harris (2003), and Straus (2004).

33. For aggression by female Milne-Edwards sifakas see Tecot et al. (2013).

34. For more information on red-winged blackbird aggression see Yasukawa and Searcy (1982).

35. For results examining male suicide in red-backed spiders see Pruitt and Riechert (2012).

CHAPTER 6. PEACOCKS, LIONS, AND MEN

1. For specific information on the northern quoll see Oakwood (2000).

2. Details on the preference of lions for male kudu can be found in Owen-Smith (1993).

3. More information on the death rates of human males and females can be found in Hawton (2000) and in the National Adolescent Health Center's 2006 Fact Sheet on Mortality: Adolescents and Young Adults, http://nahic.ucsf.edu/downloads/Mortality.pdf (accessed February 19, 2014).

4. See Kruger and Nesse (2006) for additional information on mortality rates in adult men.

5. For more details on male suicide in Italy during 2012 see Povoledo and Carvajal (2012).

6. For additional information on current research on male sensitivity see Simon and Barrett (2010) and Mac an Ghaill and Haywood (2012).

7. For gift giving in spiders see LeBas and Hockham (2005) and Albo et al. (2011).

8. For more information on cricket spermatophores see Wedell and Ritchie (2004).

9. See Le Boeuf and Mesnick (1991) for information on elephant seal mating behavior.

10. For specific information on the benefits of supporting alpha males in chimpanzees see Duffy et al. (2007).

11. For details on coalition formation in male lions see Funston et al. (2003) and Vanderwaal et al. (2009).

12. See Huffard et al. (2008) for information on beta male Indonesian octopuses.

13. For more information on friendships in Assamese macaques see Ostner et al. (2013).

14. Details on gamma males in marine isopods can be found in Shuster and Wade (1991).

15. Additional information on harassment by guppies can be found in Valero et al. (2008), Darden and Croft (2008), and Darden and Watts (2011).

16. For more information on homosexual behavior in insects see Scharf and Martin (2013).

17. For information on harassment of divers by sea turtles see Bowen (2007).

18. Details about water strider mating practices can be found in Arnqvist and Rowe (2002) and Han and Jablonski (2009).

19. See the Rape, Abuse and Incest National Network (http://www.rainn.org/get-information/statistics/sexual-assault-victims) for specifics on sexual assault rates in humans.

20. For specific information on diving beetles see Bergsten and Miller (2007).

21. Additional information on sexual coercion in orangutans can be found in Knott et al. (2010).

22. For brown bat sexual behavior see Thomas et al. (1979).

23. For physical assault in primates see Smuts and Smuts (1993).

24. See Brouwer et al. (1998) for more information.

25. For nonpaternity rates in Trinidad see Flinn (1988).

26. For more information on mating in honeybees see Koeniger and Koeniger (2007).

27. For specific details on copulatory plugs in ring-tailed lemurs see Parga (2003).

28. See Koprowski (1992) for additional information on copulatory plugs in tree squirrels.

29. For more information on chemical and toxic effects of sperm in fruit files see Johnstone and Keller (2000).

30. For the chemical composition of human sperm and its effect on mood see Gallup et al. (2002).

31. See Aisenberg et al. (2007) for details on wolf spider mating behavior.

32. For specific information on the reproductive behavior of sea horses see Mattle and Wilson (2009).

33. For mate guarding in mandrills see Setchell and Wickings (2006).

34. For chimpanzee reproductive success and male chimpanzee preference for older females see Muller et al. (2006).

35. For a review of sperm allocation by males see Wedell et al. (2002).

36. For rates of domestic violence and injury by gender see Harris (2003) and Straus (2004).

37. For more information on aggressive interactions in olive baboons see Smuts (1985) and Smuts and Smuts (1993).

38. For prairie vole aggressive behavior see Getz et al. (1981).

39. For specific information on male harassment in sheep see Réale et al. (1996).

40. See Le Galliard et al. (2005) for details on experiments altering population sex ratios in lizards.

CHAPTER 7. ARE WE MATING OR DATING?

1. For information on the mating behavior of thirteen-lined ground squirrels see Schwagmeyer and Woontner (1986).

2. For information on one-night stands and female preferences see Stewart et al. (2000), Provost et al. (2008), and Garcia et al. (2012).

3. For information on one-night stands and male preferences see Schmitt et al. (2001), Stewart et al. (2000) and Garcia et al. (2012)

4. For details on the "benefits" of short-term mating strategies in females see Buss and Schmitt (1993).

5. See Campbell (2008) and Paul and Hayes (2002) for information on the sexual satisfaction of women in short-term mating situations.

6. For specifics on social isolation and female short-term mating strategies see Sacco et al. (2012).

7. For more on crabs in ladybugs see Webberley et al. (2006).

8. See Stewart et al. (2000), Provost et al. (2008), and Garcia et al. (2012) for costs of short-term mating strategies.

9. For additional information on discrepancies in the reporting of sexual conquests see Alexander and Fisher (2003).

10. For complaints and regrets of both males and females regarding short-term mating opportunities see Campbell (2008), Hughes and Kruger (2011).

11. For details on albatross courtship rituals see Pickering and Berrow (2001).

12. See Speck and Witthoft (1947) for information on and a historical account of barnacle geese.

13. For additional information on dating in barnacle geese see Van Der Jeugd and Blaakmeer (2001).

14. For studies on personality and mating success in tits see Dingemanse et al. (2004).

15. For more details on how online dating sites make matches see Hitsch et al. (2006) and Epstein (2007).

16. For compatibility and reproductive success in cockatiels see Spoon et al. (2006).

17. For specifics on the benefits of masturbation in human males see Greening (2007).

18. For additional information on human male orgasm see Robbins and Jensen (1978) and Dunn and Trost (1989).

19. See Georgiadis et al. (2012) for details on brain activity in rats.

20. For more information on some current hypotheses regarding the female orgasm see Lloyd (2005) and Lee (2013).

21. For rates of orgasm in human females see Barr et al. (2002).

22. For a description of orgasm in macaques see Troisi and Carosi (1998) and Chevalier-Skolnikoff (1974).

23. For disruption of the sexual pleasure cycle see Georgiadis et al. (2012).

24. See Petersson and Jarvi (2001) for more information on fake orgasms in fish.

25. For more on faking orgasms in both men and women see Muehlenhard and Shippee (2010).

26. For more on human orgasms see Grupper (2005).

CHAPTER 8. THE THREE CS: COMMUNICATION, COOPERATION, AND COMPROMISE

1. For specific information on canine play see Bekoff (1995).

2. See Palagi (2008) for details on bonobo play.

3. For additional information on male treefrogs see Reichert (2010).

4. For additional information on how human males process voices see Sokhi et al. (2005).

5. See Price (2012) for specific information on how humans process language.

6. Information on tropical boubou communication can be found in Grafe and Bitz (2003).

7. For a review on the effect of human noise and disturbance on animals see Francis and Barber (2013).

8. For more on siamang mating behavior see Haimoff (1981) and Mitani (1985).

9. For specific information on experiments with honeybees see Klein et al. (2010).

10. For additional details on sleep deprivation and communication see Harrison and Horne (1997, 2000).

11. See Logue et al. (2008) for specifics on black-bellied wrens.

12. For more on purple-crowned fairywrens see Hall and Peters (2008).

13. For details on division of household labor see Shelton and John (1996) and Kamo (2000).

14. For additional information on division of chores in gay and lesbian couples see Perlesz et al. (2010).

15. See Carter and Wilkinson (2013) for more on reciprocal altruism in vampire bats.

16. A video documenting the experiments with Vulcan and Virgil can be viewed at http://www.youtube.com/watch?v=ePgC91kcmN0 (accessed February 27, 2014).

17. For more information on equal work for equal "pay" in capuchins see Brosnan and de Waal (2003).

18. See Milinski et al. (1990) for stickleback sensitivity to cooperation.

19. For additional details on shell trading in hermit crabs see Rotjan et al. (2010).

20. For voting and decision making in animals see Conradt and Roper (2005) and Conradt and List (2009).

21. Specifics on despotic societies can be found in Vehrencamp (1983).

CHAPTER 9. GETTING CUCKOLDED

1. For specific details on infidelity rates in humans see Buss and Shackelford (1997).

2. See Buss and Shackelford (1997) for a review.

3. For more on the California mouse see Ribble (1991) and Becker et al. (2012).

4. For additional information on the mating behavior of black vultures see Decker et al. (1993).

5. For details on average nonpaternity rates in humans see Cerda-Flores et al. (1999) and Voracek et al. (2009).

6. See Kraaijeveld et al. (2007) for specific information on extrapair paternity in swans.

7. For inbreeding in the Habsburg dynasty see Alvarez et al. (2009).

8. For more on extrapair copulations in the Lariang tarsier see Driller et al. (2009).

9. For swift fox mating behavior see Kitchen et al. (2006).

10. For a review on polyandry in human societies see Starkweather and Hames (2012).

11. For details on prairie vole fidelity and the role of vasopressin see Winslow et al. (1993).

12. For information on genetic influences on vasopressin and bonding see Walum et al. (2008).

13. For the average American home size, see the census figures at http://www.census.gov/const/C25Ann/sftotalmedavgsqft.pdf (accessed February 25, 2014).

14. Additional information on angelfish behavior can be found at http://www.sta.uwi.edu/fst/lifesciences/documents/Pomacanthus_paru.pdf (accessed February 25, 2013).

15. For more specifics on success of human relationships see Gottman and Notarius (2000).

16. For information on the mating seasons of bald eagles see http://www.fws.gov/midwest/eagle/pdf/nest-seasons.pdf (accessed February 25, 2014).

17. For mate support in Bewick's swans see Scott (1980).

18. For details about divorce in common terns see Gonzalez-Solis et al. (1999).

19. For more on infidelity in waved albatrosses see Huyvaert et al. (2006).

20. See Copen et al. (2012) for statistics on the divorce rate in humans in the United States between 2006 and 2010.

21. For information on divorce in common terns see Gonzalez-Solis et al. (1999).

22. For divorce in Hadza society see Marlowe (2004).

23. See Maness and Anderson (2008) for details on mate switching in boobies.

REFERENCES

Aisenberg, A., C. Viera, and F. G. Costa. 2007. Daring females, devoted males, and reversed sexual size dimorphism in the sand-dwelling spider *Allocosa brasiliensis* (Araneae, Lycosidae). *Behavioral Ecology and Sociobiology* 62 (1): 29–35.

Alain, J., S. Nusslé, A. Britschgi, G. Emanon, R. Müller, and C. Wedekind. 2007. Male dominance linked to size and age, but not to "good genes" in brown trout (*Salmo trutta*). *BMC Evolutionary Biology* 7: 207.

Alatalo, R. V., A. Lundberg, and C. Glynn. 1986. Female pied flycatchers choose territory quality and not male characteristics. *Nature* 323 (6084): 152–153.

Alberts, S. C., and C. L. Fitzpatrick. 2012. Paternal care and the evolution of exaggerated sexual swellings in primates. *Behavioral Ecology* 23 (4): 699–706.

Albo, M. J., G. Winther, C. Tuni, S. Toft, and T. Bilde. 2011. Worthless donations: male deception and female counter play in a nuptial gift-giving spider. *BMC Evolutionary Biology* 11.

Alexander, M. G., and T. D. Fisher. 2003. Truth and consequences: using the bogus pipeline to examine sex differences in self-reported sexuality. *Journal of Sex Research* 40 (1): 27–35.

Alonso-Alvarez, C., L. Perez-Rodriguez, M. E. Ferrero, E. G. de-Blas, F. Casas, and F. Mougeot. 2012. Adjustment of female reproductive investment according to male carotenoid-based ornamentation in a gallinaceous bird. *Behavioral Ecology and Sociobiology* 66 (5): 731–742.

Alvarez, G., F. C. Ceballos, and C. Quinteiro. 2009. The role of inbreeding in the extinction of a European royal dynasty. *PLOS ONE* 4 (4).

Andersson, M. 1982. Female choice selects for extreme tail length in a widowbird. *Nature* 299 (5886): 818–820.

Archer, J. 2000. Sex differences in aggression between heterosexual partners: a meta-analytic review. *Psychological Bulletin* 126 (5): 651–680.

———. 2006. Testosterone and human aggression: an evaluation of the challenge hypothesis. *Neuroscience and Biobehavioral Reviews* 30 (3): 319–345.

Arnqvist, G., and L. Rowe. 2002. Correlated evolution of male and female morphologies in water striders. *Evolution* 56 (5): 936–947.

Artiss, T., and K. Martin. 1995. Male vigilance in white-tailed ptarmigan, *Lagopus leucurus*: mate guarding or predator detection? *Animal Behaviour* 49 (5): 1249–1258.

Asprey, R. B. 2001. *The Rise of Napoleon Bonaparte*. New York: Basic Books.

Backwell, P. R. Y., J. H. Christy, S. R. Telford, M. D. Jennions, and N. I. Passmore. 2000. Dishonest signalling in a fiddler crab. *Proceedings of the Royal Society B* (1444): 719–724.

Barr, A., A. Bryan, and D. T. Kenrick. 2002. Sexual peak: socially shared cognitions about desire, frequency, and satisfaction in men and women. *Personal Relationships* 9 (3): 287–299.

Beck, D. M., M. A. Pinsk, and S. Kastner. 2005. Symmetry perception in humans and macaques. *Trends in Cognitive Sciences* 9 (9): 405–406.

Becker, E. A., S. Petruno, and C. A. Marler. 2012. A comparison of scent marking between a monogamous and promiscuous species of *Peromyscus*: pair bonded males do not advertise to novel females. *PLOS ONE* 7 (2).

Bee, M. A., S. A. Perrill, and P. C. Owen. 2000. Male green frogs lower the pitch of acoustic signals in defense of territories: a possible dishonest signal of size? *Behavioral Ecology* 11 (2): 169–177.

Bekoff, M. 1995. Play signals as punctuation: the structure of social play in canids. *Behaviour* 132: 419–429.

Belluscio, L., G. Koentges, R. Axel, and C. Dulac. 1999. A map of pheromone receptor activation in the mammalian brain. *Cell* 97: 209–220.

Bergsten, J., and K. B. Miller. 2007. Phylogeny of diving beetles reveals a coevolutionary arms race between the sexes. *PLOS ONE* 2: e522.

Berry, D. S. 2000. Attractiveness, attraction, and sexual selection: evolutionary perspectives on the form and function of physical attractiveness. *Advances in Experimental Social Psychology* 32: 273–342.

Beydoun, H., and A. F. Saftlas. 2004. Association of human leukocyte antigen sharing with recurrent spontaneous abortions. *Tissue Antigens* 65: 123–135.

Birkhead, T. R., and A. P. Moller. 1993. Sexual selection and the temporal separation of reproductive events: sperm storage data from reptiles, birds, and mammals. *Biological Journal of the Linnean Society* 50 (4): 295–311.

Boogert, N. J., T. W. Fawcett, and L. Lefebvre. 2011. Mate choice for cognitive traits: a review of the evidence in nonhuman vertebrates. *Behavioral Ecology* 22 (3): 447–459.

Borries, C., K. Launhardt, C. Epplen, J. T. Epplen, and P. Winkler. 1999. DNA analyses support the hypothesis that infanticide is adaptive in langur monkeys. *Proceedings of the Royal Society B* 266 (1422): 901–904.

Boulton, R. A., and C. Ross. 2013. Measuring facial symmetry in the wild: a case study in Olive Baboons (*Papio anubis*). *Behavioral Ecology and Sociobiology* 67 (4): 699–707.

Bowen, B. W. 2007. Sexual harassment by a male green turtle (*Chelonia mydas*).

Marine Turtle Newsletter. http://www.seaturtle.org/mtn/archives/mtn117/mtn117p10.shtml?nocount (accessed February 20, 2014).

Bowman, E. A. 2010. An explanation for the shape of the human penis. *Archives of Sexual Behavior* 39 (2): 216.

Brand, R. J., A. Bonatsos, R. D'Orazio, and H. DeShong. 2012. What is beautiful is good, even online: correlations between photo attractiveness and text attractiveness in men's online dating profiles. *Computers in Human Behavior* 28 (1): 166–170.

Brand, R. J., C. M. Markey, A. Mills, and S. D. Hodges. 2007. Sex differences in self-reported infidelity and its correlates. *Sex Roles* 57 (1–2): 101–109.

Brennan, P. A., and K. M. Kendrick. 2006. Mammalian social odours: attraction and individual recognition. *Philosophical Transactions of the Royal Society B* 361 (1476): 2061–2078.

Brennan, P. L. R., R. O. Prum, K. G McCracken, M. D. Sorenson, R. E. Wilson, and T. R. Birkhead. 2007. Coevolution of male and female genital morphology in waterfowl. *PLOS ONE* 2 (5): e418.

Breuer, T., A. M. Robbins, C. Boesch, and M. M. Robbins. 2012. Phenotypic correlates of male reproductive success in western gorillas. *Journal of Human Evolution* 62 (4): 466–472.

Brosnan, S. F., and F. B. M. de Waal. 2003. Monkeys reject unequal pay. *Nature* 425 (6955): 297–299.

Brouwer, E. C., B. M. Harris, and T. Sonomi. 1998. *Gender Analysis in Papua New Guinea*. Washington, DC: World Bank.

Burbank, V. K. 1987. Female aggression in cross-cultural perspective. *Cross-Cultural Research* 21 (1–4): 70–100.

Burley, N. 1985. Leg-band color and mortality patterns in captive breeding populations of zebra finches. *Auk* 102 (3): 647–651.

———. 1986a. Sex-ratio manipulation in color-banded populations of zebra finches. *Evolution* 40 (6): 1191–1206.

———. 1986b. Sexual selection for aesthetic traits in species with biparental care. *American Naturalist* 127 (4): 415–445.

Buss, D. M. 1989. Sex-differences in human mate preferences: evolutionary hypothesis tested in 37 cultures. *Behavioral and Brain Sciences* 12 (1): 1–14.

Buss, D. M., and D. P. Schmitt. 1993. Sexual strategies theory: an evolutionary perspective on human mating. *Psychological Review* 100 (2): 204–232.

Buss, D. M., and T. K. Shackelford. 1997. Susceptibility to infidelity in the first year of marriage. *Journal of Research in Personality* 31 (2): 193–221.

Campbell, Anne. 2008. The morning after the night before. *Human Nature* 19 (2): 157–173.

Carpenter, C. C. 1977. Communication and displays of snakes. *American Zoologist* 17 (1): 217–223.

———. 1986. An inventory of combat rituals in snakes. Smithsonian Herpetological Information Service: 1–18.

Carrieri, V., and M. De Paola. 2012. Height and subjective well-being in Italy. *Economics and Human Biology* 10 (3): 289–298.

Carter, G, G., and G. S. Wilkinson. 2013. Food sharing in vampire bats: reciprocal help predicts donations more than relatedness or harassment. *Proceedings of the Royal Society B* 280 (1753): 1–6.

Catoni, C., A. Peters, and H. M. Schaefer. 2009. Dietary flavonoids enhance conspicuousness of a melanin-based trait in male blackcaps but not of the female homologous trait or of sexually monochromatic traits. *Journal of Evolutionary Biology* 22 (8): 1649–1657.

Cerda-Flores, R. M., S. A. Barton, L. F. Marty-Gonzalez, F. Rivas, and R. Chakraborty. 1999. Estimation of nonpaternity in the Mexican population of Nuevo Leon: a validation study with blood group markers. *American Journal of Physical Anthropology* 109 (3): 281–293.

Chamorro-Florescano, I. A., M. E. Favila, and R. Macias-Ordonez. 2011. Ownership, size, and reproductive status affect the outcome of food ball contests in a dung roller beetle: when do enemies share? *Evolutionary Ecology* 25 (2): 277–289.

Chapman, D. D., M. S. Shivji, E. Louis, J. Sommer, H. Fletcher, and P. A. Prodohl. 2007. Virgin birth in a hammerhead shark. *Biology Letters* 3 (4): 425–427.

Chase, I. D., D. Bartolomeo, L. A. Dugatkin. 1994. Aggressive interactions and inter-contest interval: how long do winners keep winning? *Animal Behaviour* 48: 393–400.

Chen, M. K., V. Lakshminarayanan, and L. R. Santos. 2006. How basic are behavioral biases? Evidence from capuchin monkey trading behavior. *Journal of Political Economy* 114 (3): 517–537.

Chesh, A. S., K. E. Mabry, B. Keane, D. A. Noe, and N. G. Solomon. 2012. Are body mass and parasite load related to social partnerships and mating in *Microtus ochrogaster*? *Journal of Mammalogy* 93 (1): 229–238.

Chevalier-Skolnikoff, S. 1974. Male-female, female-female, and male-male sexual behavior in stumptail monkey, with special attention to female orgasm. *Archives of Sexual Behavior* 3 (2): 95–116.

Clutton-Brock, T., and K. McAuliffe. 2009. Female mate choice in mammals. *Quarterly Review of Biology* 84 (1): 3–27.

Connor, R. C., R. A. Smolker, and A. F. Richards. 1992. Two levels of alliance formation among male bottlenose dolphins (*Tursiops* sp). *Proceedings of the National Academy of Sciences of the United States of America* 89 (3): 987–990.

Conradt, L., and C. List. 2009. Group decisions in humans and animals: a survey. *Philosophical Transactions of the Royal Society B* 364 (1518): 719–742.

Conradt, L., and T. J. Roper. 2005. Consensus decision making in animals. *Trends in Ecology and Evolution* 20 (8): 449–456.

Constant, N., D. Valbuena, and C. C. Rittschof. 2011. Male contest investment changes with male body size but not female quality in the spider *Nephila clavipes*. *Behavioural Processes* 87 (2): 218–223.

Copen, C. E., K. Daniels, J. Vespa, and W. D. Mosher. 2012. First marriages in the United States: data from the 2006–2010 national survey of family growth. *National Health Statistics Reports* (49): 1–21.

Courtiol, A., S. Picq, B. Godelle, M. Raymond, and J. B. Ferdy. 2010. From preferred to actual mate characteristics: the case of human body shape. *PLOS ONE* 5 (9): e13010.

Cronk, L., and B. Dunham. 2007. Amounts spent on engagement rings reflect aspects of male and female mate quality. *Human Nature* 18 (4): 329–333.

Currey, J. D., T. Landete-Castillejos, J. Estevez, F. Ceacero, A. Olguin, A. Garcia, and L. Gallego. 2009. The mechanical properties of red deer antler bone when used in fighting. *Journal of Experimental Biology* 212 (24): 3985–3993.

Cuthill, I. C., S. Hunt, C. Cleary, and C. Clark. 1997. Colour bands, dominance, and body mass regulation in male zebra finches (*Taeniopygia guttata*). *Proceedings of the Royal Society B* 264 (1384): 1093–1099.

Danel, D. P., P. Fedurek, V. Coetzee, I. D. Stephen, N. Nowak, M. Stirrat, D. I. Perrett, and T. K. Saxton. 2012. A cross-cultural comparison of population-specific face shape preferences (*Homo sapiens*). *Ethology* 118 (12): 1173–1181.

Darden, S. K, and D. P. Croft. 2008. Male harassment drives females to alter habitat use and leads to segregation of the sexes. *Biology Letters* 4 (5): 449–451.

Darden, S. K., and L. Watts. 2011. Male sexual harassment alters female social behaviour towards other females. *Biology Letters* 8 (2): 186–188.

de Solla, S. R., M. Portelli, H. Spiro, and R. J. Brooks. 2001. Penis displays of common snapping turtles (*Chelydra serpentina*) in response to handling: defensive or displacement behaviour? *Chelonian Conservation and Biology* 4: 187–189.

Decker, M. D., P. G. Parker, D. J. Minchella, and K. N. Rabenold. 1993. Monogamy in black vultures: genetic evidence from DNA fingerprinting. *Behavioral Ecology* 4 (1): 29–35.

Deinert, E. I. 2003. Sexual selection in *Heliconius hewitsoni*, a pupal mating butterfly. In *Ecology and Evolution Taking Flight: Butterflies as Model Study Systems*, edited by C. L. Boggs, P. R. Ehrlich, and W. B. Watt. Chicago: University of Chicago Press.

del Cerro, S., S. Merino, J. Martinez-de la Puente, E. Lobato, R. Ruiz-de-Castaneda, J. Rivero-de Aguilar, J. Martinez, J. Morales, G. Tomas, and J. Moreno. 2010. Carotenoid-based plumage colouration is associated with blood parasite richness and stress protein levels in blue tits (*Cyanistes caeruleus*). *Oecologia* 162 (4): 825–835.

Diamond, J. M. 1986. Ethnic differences: variation in human testis size. *Nature* 320 (6062): 488–489.

Dingemanse, N. J., C. Both, P. J. Drent, and J. M. Tinbergen. 2004. Fitness consequences of avian personalities in a fluctuating environment. *Proceedings of the Royal Society B* 271 (1541): 847–852.

Dixson, B. J., and R. C. Brooks. 2013. The role of facial hair in women's perceptions of men's attractiveness, health, masculinity, and parenting abilities. *Evolution and Human Behavior* 34 (3): 236–241.

Dixson, B. J., A. F. Dixson, P. J. Bishop, and A. Parish. 2010a. Human physique and sexual attractiveness in men and women: a New Zealand–US comparative study. *Archives of Sexual Behavior* 39 (3): 798–806.

Dixson, B. J., G. M. Grimshaw, W. L. Linklater, and A. F. Dixson. 2010b. Watching the hourglass: eye-tracking reveals men's appreciation of the female form. *Human Nature* 21 (4): 355–370.

———. 2011a. Eye-tracking of men's preferences for waist-to-hip ratio and breast size of women. *Archives of Sexual Behavior* 40 (1): 43–50.

Dixson, B. J., J. C. Tam, and M. Awasthy. 2013. Do women's preferences for men's facial hair change with reproductive status? *Behavioral Ecology* 24 (3): 708–716.

Dixson, B. J., and P. L. Vasey. 2012. Beards augment perceptions of men's age, social status, and aggressiveness, but not attractiveness. *Behavioral Ecology* 23 (3): 481–490.

Dixson, B. J., P. L. Vasey, K. Sagata, N. Sibanda, W. L. Linklater, and A. F. Dixson. 2011b. Men's preferences for women's breast morphology in New Zealand, Samoa, and Papua New Guinea. *Archives of Sexual Behavior* 40 (6): 1271–1279.

Doucet, S. M., and R. Montgomerie. 2003. Multiple sexual ornaments in satin bowerbirds: ultraviolet plumage and bowers signal different aspects of male quality. *Behavioral Ecology* 14 (4): 503–509.

Doutrelant, C., A. Gregoire, N. Grnac, D. Gomez, M. M. Lambrechts, and P.

Perret. 2008. Female coloration indicates female reproductive capacity in blue tits. *Journal of Evolutionary Biology* 21 (1): 226–233.

Driller, C., D. Perwitasari-Farajallah, H. Zischler, and S. Merker. 2009. The social system of Lariang tarsiers (*Tarsius lariang*) as revealed by genetic analyses. *International Journal of Primatology* 30 (2): 267–281.

Duffy, K. G., R. W. Wrangham, and J. B. Silk. 2007. Male chimpanzees exchange political support for mating opportunities. *Current Biology* 17 (15): R586–R587.

Dunn, M. E, and J. E. Trost. 1989. Male multiple orgasms: a descriptive study. *Archives of Sexual Behavior* 18 (5):377–387.

Dunn, P. O., and A. Cockburn. 1999. Extrapair mate choice and honest signaling in cooperatively breeding superb fairy-wrens. *Evolution* 53 (3): 938–946.

Economist. 2003. The right to be beautiful. *Economist*: May 22, 2003.

Elwood, R. W., R. M. E. Pothanikat, and M. Briffa. 2006. Honest and dishonest displays, motivational state, and subsequent decisions in hermit crab shell fights. *Animal Behaviour* 72: 853–859.

Emlen, S. T., and P. H. Wrege. 2004. Size dimorphism, intrasexual competition, and sexual selection in wattled jacana (*Jacana jacana*), a sex-role-reversed shorebird in Panama. *Auk* 121 (2): 391–403.

Epstein, R. 2007. The truth about online dating. *Scientific American Mind* 18: 28–35.

Evans, S., N. Neave, and D. Wakelin. 2006. Relationships between vocal characteristics and body size and shape in human males: an evolutionary explanation for a deep male voice. *Biological Psychology* 72 (2): 160–163.

Fan J., W. Dai, F. Liu, and J. Wu. 2005. Visual perception of male body attractiveness. *Proceedings of the Royal Society B* 272: 219–226.

Farmer, D. C., and C. J. Barnard. 2000. Fluctuating asymmetry and sperm transfer in male decorated field crickets (*Gryllodes sigillatus*). *Behavioral Ecology and Sociobiology* 47: 287–292.

Fedy, B. C., and T. E. Martin. 2009. Male songbirds provide indirect parental care by guarding females during incubation. *Behavioral Ecology* 20 (5): 1034–1038.

Firman, R. C., L. W. Simmons, J. M. Cummins, and P. L. Matson. 2003. Are body fluctuating asymmetry and the ratio of 2nd to 4th digit length reliable predictors of semen quality? *Human Reproduction* 18 (4): 808–812.

Flinn, M. V. 1988. Mate guarding in a Caribbean village. *Ethology and Sociobiology* 9 (1): 1–28.

Francis, C. D., and J. R. Barber. 2013. A framework for understanding noise impacts on wildlife: an urgent conservation priority. *Frontiers in Ecology and the Environment* 11 (6): 305–313.

Funston, P. J., M. G. L. Mills, P. R. K. Richardson, and A. S. van Jaarsveld. 2003. Reduced dispersal and opportunistic territory acquisition in male lions (*Panthera leo*). *Journal of Zoology* 259: 131–142.

Fuxjager, M. J., G. Mast, E. A. Becker, and C. A. Marler. 2009. The "home advantage" is necessary for a full winner effect and changes in post-encounter testosterone. *Hormones and Behavior* 56 (2): 214–219.

Gadagkar, R. 2003. Is the peacock merely beautiful or also honest? *Current Science* 85 (7): 1012–1020.

Galeotti, P., R. Sacchi, and V. Vicario. 2005. Fluctuating asymmetry in body traits increases predation risks: tawny owl selection against asymmetric woodmice. *Evolutionary Ecology* 19 (4): 405–418.

Gallup, G. G., R. L. Burch, and T. J. B. Mitchell. 2006. Semen displacement as a sperm competition strategy: multiple mating, self-semen displacement, and timing of in-pair copulations. *Human Nature* 17 (3): 253–264.

Gallup, G. G., R. L. Burch, M. L. Zappieri, R. A. Parvez, M. L. Stockwell, and J. A. Davis. 2003. The human penis as a semen displacement device. *Evolution and Human Behavior* 24 (4): 277–289.

Gallup, G. G., R. L. Burch, and S. Platek. 2002. Does semen contain antidepressant properties? *Archives of Sexual Behavior* 39: 289–291.

Galvan, I., C. Alonso-Alvarez, and J. J. Negro. 2012. Relationships between hair melanization, glutathione levels, and senescence in wild boars. *Physiological and Biochemical Zoology* 85 (4): 332–347.

Garcia, J. R., C. Reiber, S. G. Massey, and A. M. Merriwether. 2012. Sexual hookup culture: a review. *Review of General Psychology* 16 (2): 161–176.

Garcia-Navas, V., E. S. Ferrer, and J. J. Sanz. 2012. Plumage yellowness predicts foraging ability in the blue tit Cyanistes caeruleus. *Biological Journal of the Linnean Society* 106 (2): 418–429.

Garver-Apgar, C., S. W. Gangestad, R. Thornhill, R. D. Miller, and J. J. Olp. 2006. Major histocompatibility complex alleles, sexual responsivity, and unfaithfulness in romantic couples. *Psychological Science* 7 (10): 830–835.

Gates, H. 2001. Footloose in Fujian: economic correlates of footbinding. *Comparative Studies in Society and History* 43 (1): 130–148.

Gemeno, C., and J. Claramunt. 2006. Sexual approach in the praying mantid *Mantis religiosa* (L.). *Journal of Insect Behavior* 19 (6): 731–740.

Georgiadis, J. R., M. L. Kringelbach, and J. G. Pfaus. 2012. Sex for fun: a synthesis of human and animal neurobiology. *Nature Reviews Urology* 9 (9): 486–498.

Gersick, A. S., N. Snyder-Mackler, and D. J. White. 2012. Ontogeny of social skills: social complexity improves mating and competitive strategies in male brown-headed cowbirds. *Animal Behaviour* 83 (5): 1171–1177.

Getz, L. L., C. S. Carter, and L. Gavish. 1981. The mating system of the prairie vole, *Microtus ochrogaster*: field and laboratory evidence for pair-bonding. *Behavioral Ecology and Sociobiology* 8 (3): 189–194.

Gomendio, M., J. Cassinello, and E. R. S. Roldan. 2000. A comparative study of ejaculate traits in three endangered ungulates with different levels of inbreeding: fluctuating asymmetry as an indicator of reproductive and genetic stress. *Proceedings of the Royal Society B* 267: 875–882.

Gomes, C. M., and C. Boesch. 2009. Wild chimpanzees exchange meat for sex on a long-term basis. *PLOS ONE* 4 (4).

Gonzalez-Solis, J., P. H. Becker, and H. Wendeln. 1999. Divorce and asynchronous arrival in common terns, *Sterna hirundo*. *Animal Behaviour* 58: 1123–1129.

Gosselin, C. 1984. Hair loss, personality, and attitudes. *Personality and Individual Differences* 5 (3): 365–369.

Gottman, J. M., and C. I. Notarius. 2000. Decade review: observing marital interaction. *Journal of Marriage and the Family* 62 (4): 927–947.

Grafe, T. U., and J. H. Bitz. 2003. The functions of duetting in the tropical boubou (*Laniarius aethiopicus*): experimental evidence for territorial defence. *Animal Behaviour* 68 (1): 193–201.

Grafe, T. U., J. H. Bitz, and M. Wink. 2004. Song repertoire and duetting behaviour of the tropical boubou, *Laniarius aethiopicus*. *Animal Behaviour* 68 (1): 181–191.

Greening, D. J. 2007. Frequent ejaculation: a pilot study of changes in sperm DNA damage and semen parameters using daily ejaculation. *Fertility and Sterility* 88: S19–S20.

Greiling, H., and D. M. Buss. 2000. Women's sexual strategies: the hidden dimension of extra-pair mating. *Personality and Individual Differences* 28 (5): 929–963.

Groscolas, R. 1986. Changes in body mass, body temperature, and plasma fuel levels during the natural breeding fast in male and female emperor penguins *Aptenodytes forsteri*. *Journal of Comparative Physiology B* 156 (4): 521–527.

Grupper, J. (director). 2005. *Anatomy of Sex* [documentary]. United States: Discovery Channel.

Gumert, M. D. 2007. Payment for sex in a macaque mating market. *Animal Behaviour* 74 (6): 1655–1667.

Gunderson, E. K. 1965. Body size, self-evaluation, and military effectiveness. *Journal of Personality and Social Psychology* 2 (6): 902–906.

Haimoff, E. H. 1981. Video analysis of siamang (*Hylobates syndactylus*) songs. *Behaviour* 76: 128–151.

Hale, M. L., M. H. Verduijn, A. P. Moller, K. Wolff, and M. Petrie. 2009. Is the peacock's train an honest signal of genetic quality at the major histocompatibility complex? *Journal of Evolutionary Biology* 22 (6): 1284–1294.

Hall, M. L., and A. Peters. 2008. Coordination between the sexes for territorial defence in a duetting fairy-wren. *Animal Behaviour* 76: 65–73.

Hamilton, J. B. 1942. Male hormone stimulation is a prerequisite and an incitant in common baldness. *American Journal of Anatomy* 71: 451–480.

Hampton, K. J., K. A. Hughes, and A. E. Houde. 2009. The allure of the distinctive: reduced sexual responsiveness of female guppies to "redundant" male colour patterns. *Ethology* 115 (5): 475–481.

Han, C. S., and P. G. Jablonski. 2009. Female genitalia concealment promotes intimate male courtship in a water strider. *PLOS ONE* 4: e5793.

Hansen K. 2007. Hair or bare? The history of American women and hair removal, 1914–1934 (unpublished senior thesis in American Studies, Barnard College, Columbia University), http://history.barnard.edu/sites/default/files/inline/kirstenhansenthesis.pdf (accessed February 19, 2014).

Harano, T., N. Sato, and T. Miyatake. 2012. Effects of female and male size on female mating and remating decisions in a bean beetle. *Journal of Ethology* 30 (3): 337–343.

Harris, C. R. 2003. A review of sex differences in sexual jealousy, including self-report data, psychophysiological responses, interpersonal violence, and morbid jealousy. *Personality and Social Psychology Review* 7 (2): 102–128.

Harrison, Y., and J. A. Horne. 1997. Sleep deprivation affects speech. *Sleep* 20 (10): 871–878.

———. 2000. The impact of sleep deprivation on decision making: a review. *Journal of Experimental Psychology: Applied* 6 (3): 236.

Havlicek, J., and S. C. Roberts. 2009. MHC-correlated mate choice in humans: a review. *Psychoneuroendocrinology* 34: 497–512.

Hawton, K. 2000. Sex and suicide: gender differences in suicidal behaviour. *British Journal of Psychiatry* 177 (6): 484–485.

Heffner, L. J. 2004. Advanced maternal age: how old is too old? *New England Journal of Medicine* 351 (19): 1927–1929.

Hendrie, C. A., and G. Brewer. 2012. Evidence to suggest that teeth act as human ornament displays signalling mate quality. *PLOS ONE* 7 (7).

Hill, G. E., and K. J. McGraw. 2003. Melanin, nutrition, and the lion's mane. *Science* 299 (5607): 660.

Hill, R., and R. Barton. 2005. Psychology: red enhances human performance in contests. *Nature* 435 (7040): 293.

Hindwood, K. A. 1948. The use of flower petals in courtship displays. *Emu* 47: 389–391.

Hinsz, V. B., C. J. Stoesser, and D. C. Matz. 2013. The Intermingling of social and evolutionary psychology influences on hair color preferences. *Current Psychology* 32 (2): 136–149.

Hitsch, G. J., A. Hortacsu, and D. Ariely. 2006. What makes you click? Mate preferences and matching outcomes in online dating. MIT Sloan Paper No. 4603-06.

Hoch, J. M. 2010. Effects of crowding and wave exposure on penis morphology of the acorn barnacle, *Semibalanus balanoides*. *Marine Biology* 157 (12): 2783–2789.

Hoefler, C. D., J. A. Moore, K. T. Reynolds, and A. L. Rypstra. 2010. The effect of experience on male courtship and mating behaviors in a cellar spider. *American Midland Naturalist* 163 (2): 255–268.

Hogstad, O. 1992. Mate protection in alpha pairs of wintering willow tits, *Parus montanus*. *Animal Behaviour* 43 (2): 323–328.

Hotzy, C., and G. Arnqvist. 2009. Sperm competition favors harmful males in seed beetles. *Current Biology* 19 (5): 404–407.

Huchard, E., A. Baniel, S. Schliehe-Diecks, and P. M. Kappeler. 2013. MHC-disassortative mate choice and inbreeding avoidance in a solitary primate. *Molecular Ecology* 22 (15): 4071–4086.

Huck, U. W., E. M. Banks, and C. B. Coopersmith. 1984. Social olfaction in male brown lemmings (*Lemmus sibiricus=trimucronatus*) and collared lemmings (*Dicrostonyx groenlandicus*): 2. Discrimination of mated and unmated females. *Journal of Comparative Psychology* 98 (1): 60–65.

Huffard, C. L, R. L. Caldwell, and F. Boneka. 2008. Mating behavior of *Abdopus aculeatus* (d'Orbigny 1834) (Cephalopoda: Octopodidae) in the wild. *Marine Biology* 154 (2): 353–362.

Hughes, Susan M., and Daniel J. Kruger. 2011. Sex differences in post-coital behaviors in long- and short-term mating: an evolutionary perspective. *The Journal of Sex Research* 48 (5): 496–505.

Hugill, N., B. Fink, and N. Neave. 2010. The role of human body movements in mate selection. *Evolutionary Psychology* 8 (1): 66–89.

Hugill, N., B. Fink, N. Neave, and H. Seydel. 2009. Men's physical strength is associated with women's perceptions of their dancing ability. *Personality and Individual Differences* 47 (5): 527–530.

Hume, D. K., and R. Montgomerie. 2001. Facial attractiveness signals different aspects of "quality" in women and men. *Evolution and Human Behavior* 22 (2): 93–112.

Hunt, M. K., C. J. Nicholls, R. J. Wood, A. P. Rendon, and A. S. Gilburn. 2004. Sexual selection for symmetrical male medflies (Diptera: Tephritidae) confirmed in the field. *Biological Journal of the Linnean Society* 81 (3): 347–355.

Hunt, M. K., E. A. Roux, R. J. Wood, and A. S. Gilburn. 2002. The effect of Supra-fronto-orbital (SFO) bristle removal on male mating success in the Mediterranean fruit fly (Diptera: Tephritidae). *Florida Entomologist* 85 (1): 83–88.

Hunter, E. M, and S. L. Davis. 1998. Female Adélie penguins acquire nest material from extrapair males after engaging in extrapair copulations. *Auk* 2: 526–528.

Huyvaert, K. P., D. J. Anderson, and P. G. Parker. 2006. Mate opportunity hypothesis and extrapair paternity in waved albatrosses (*Phoebastria irrorata*). *Auk* 123 (2): 524–536.

Jacob, A., S. Nussle, A. Britschgi, G. Evanno, R. Mueller, and C. Wedekind. 2007. Male dominance linked to size and age, but not to "good genes" in brown trout (*Salmo trutta*). *BMC Evolutionary Biology* 7.

Jamieson, I. 1995. Female fish prefer to spawn in nests with eggs for reasons of mate choice copying or egg survival. *American Naturalist* 145 (5): 824–832.

Johnstone, R. A., and L. Keller. 2000. How males can gain by harming their mates: sexual conflict, seminal toxins, and the cost of mating. *American Naturalist* 156 (4): 368–377.

Kamo, Y. 2000. "He said, she said": assessing discrepancies in husbands' and wives' reports on the division of household labor. *Social Science Research* 29 (4): 459–476.

Kasumovic, M. M., D. O. Elias, D. Punzalan, A. C. Mason, and M. C. B. Andrade. 2009. Experience affects the outcome of agonistic contests without affecting the selective advantage of size. *Animal Behaviour* 77 (6): 1533–1538.

Kasumovic, M. M., D. O. Elias, S. Sivalinghem, A. C. Mason, and M. C. B. Andrade. 2010. Examination of prior contest experience and the retention of winner and loser effects. *Behavioral Ecology* 21 (2): 404–409.

Kaufmann, J. H. 1975. Field observations of social behaviour of eastern grey kangaroo, *Macropus giganteus*. *Animal Behaviour* 23 (February): 214–221.

Keagy, J, J. F. Savard, and G. Borgia. 2009. Male satin bowerbird problem-solving ability predicts mating success. *Animal Behaviour* 78 (4): 809–817.

———. 2011. Complex relationship between multiple measures of cognitive ability and male mating success in satin bowerbirds, *Ptilonorhynchus violaceus*. *Animal Behaviour* 81: 1063–1070.

Kemp, D. J., and C. Wiklund. 2001. Fighting without weaponry: a review of male-male contest competition in butterflies. *Behavioral Ecology and Sociobiology* 49 (6): 429–442.

Khan, S. A., W. J. Levine, S. D. Dobson, and J. D. Kralik. 2011. Red signals dominance in male rhesus macaques. *Psychological Science* 22 (8): 1001–1003.

Kitchen, A. M., E. M. Gese, L. P. Waits, S. M. Karki, and E. R. Schauster. 2006. Multiple breeding strategies in the swift fox, *Vulpes velox*. *Animal Behaviour* 71: 1029–1038.

Klein, B. A., A. Klein, M. K. Wray, U. G. Mueller, and T. D. Seeley. 2010. Sleep deprivation impairs precision of waggle dance signaling in honey bees. *Proceedings of the National Academy of Sciences of the United States of America* 107 (52): 22705–22709.

Knott, C. D., M. E. Thompson, R. M. Stumpf, and M. H. McIntyre. 2010. Female reproductive strategies in orangutans, evidence for female choice and counterstrategies to infanticide in a species with frequent sexual coercion. *Proceedings of the Royal Society B* 277: 105–113.

Koeniger, N., and G. Koeniger. 2007. Mating flight duration of *Apis mellifera* queens: as short as possible, as long as necessary. *Apidologie* 38: 606–611.

Koprowski, J. L.1992. Removal of copulatory plugs by female tree squirrels. *Journal of Mammalogy* 73 (3): 572–576.

Kraaijeveld, K., M. Ming, J. Komdeur, and R. A. Mulder. 2007. Offspring sex ratios in relation to mutual ornamentation and extra-pair paternity in the black swan *Cygnus atratus*. *Ibis* 149 (1): 79–85.

Kraak, S. B. M., and F. J. Weissing. 1996. Female preference for nests with many eggs: a cost-benefit analysis of female choice in fish with paternal care. *Behavioral Ecology* 7 (3): 353–361.

Kruger, D. J., and R. M. Nesse. 2006. An evolutionary life-history framework for understanding sex differences in human mortality rates. *Human nature* 17 (1): 74–97.

Lanuza, E., A. Novejarque, J. Martinez-Ricos, J. Martinez-Hernandez, C. Agustin-Pavon, and F. Martinez-Garcia. 2008. Sexual pheromones and the evolution of the reward system of the brain: the chemosensory function of the amygdala. *Brain Research Bulletin* 75 (2–4): 460–466.

Lawrence, S. E. 1992. Sexual cannibalism in the praying mantid, *Mantis religiosa*: a field study. *Animal Behaviour* 43 (4): 569–583.

LeBas, N. R., and L. R. Hockham. 2005. An invasion of cheats: the evolution of worthless nuptial gifts. *Current Biology* 15 (1): 64–67.

Le Boeuf, B. J., and S. Mesnick. 1991. Sexual behavior of male northern elephant seals: I. Lethal injuries to adult females. *Behaviour* 116 143–162.

Lee, D. J. 2013. Homology, female orgasm, and the forgotten argument of Donald Symons. *Biology and Philosophy* 28 (6): 1021–1027.

Lee, K. C., and B. Ladizinski. 2013. Hair today, gone tomorrow: the abandonment of body hair by American women. *Journal of American Dermatology* 149 (2): 208.

Leffelaar, D., and R. J. Robertson. 1985. Nest usurpation and female competition for breeding opportunities by tree swallows. *Wilson Bulletin* 97 (2): 221–224.

Le Galliard, J. F., P. S. Fitze, R. Ferrière, and J. Clobert. 2005. Sex ratio bias, male aggression, and population collapse in lizards. *Proceedings of the National Academy of Sciences of the United States of America* 102 (50): 18231–18236.

Lemaitre, J. F., S. A. Ramm, J. L. Hurst, and P. Stockley. 2012. Sperm competition roles and ejaculate investment in a promiscuous mammal. *Journal of Evolutionary Biology* 25 (6): 1216–1225.

Lemasson, A., R. A. Palombit, and R. Jubin. 2008. Friendships between males and lactating females in a free-ranging group of olive baboons (*Papio hamadryas anubis*): evidence from playback experiments. *Behavioral Ecology and Sociobiology* 62 (6): 1027–1035.

Le Page, M. 2014. Testicular time bomb: older dads' mutant sperm. *New Scientist* 2957, available online at http://www.newscientist.com/article/mg22129570.800-testicular-time-bomb-older-dads-mutant-sperm.html?full=true#.Uw9HcBzx-Pp (accessed February 27, 2014).

Lie, H. C., L. W. Simmons, and G. Rhodes. 2010. Genetic dissimilarity, genetic diversity, and mate preferences in humans. *Evolution and Human Behavior* 31: 48–58.

Ligon, J. D., R. Kimball, and M. Merola-Zwartjes. 1998. Mate choice by female red junglefowl: the issues of multiple ornaments and fluctuating asymmetry. *Animal Behaviour* 55: 41–50.

Limbourg, T., C. A. Mateman, and C. M. Lessells. 2013. Parental care and UV coloration in blue tits: opposite correlations in males and females between provisioning rate and mate's coloration. *Journal of Avian Biology* 44: 17–26.

Lindgaard, G., G. Fernandes, C. Dudekx, and J. Brown. 2006. Attention web designers: you have 50 milliseconds to make a good first impression! *Behaviour and Information Technology* 25 (2): 115–126.

Little, A. C., A. Paukner, R. A. Woodward, and S. J. Suomi. 2012. Facial asymmetry is negatively related to condition in female macaque monkeys. *Behavioral Ecology and Sociobiology* 66 (9): 1311–1318.

Little, A. C., and B. C. Jones. 2006. Attraction independent of detection suggests special mechanisms for symmetry preferences in human face perception. *Proceedings of the Royal Society B* 273 (1605): 3093–3099.

Little, A. C., B. C. Jones, and L. M. DeBruine. 2011. Facial attractiveness: evolutionary based research. *Philosophical Transactions of the Royal Society B* 366 (1571): 1638–1659.

Little, A. C., B. C. Jones, C. Waitt, B. P. Tiddeman, D. R. Feinberg, D. I. Perrett, C. L. Apicella, and F. W. Marlowe. 2008. Symmetry is related to sexual dimorphism in faces: data across culture and species. *PLOS ONE* 3 (5).

Lloyd, E. A. 2005. *The Case of the Female Orgasm: Bias in the Science of Evolution* Cambridge, MA: Harvard University Press.

Lloyd, J. E. 1984. Occurrence of aggressive mimicry in fireflies. *Florida Entomologist* 67 (3): 368–376.

Logue, D. M., C. Chalmers, and A. H. Gowland. 2008. The behavioural mechanisms underlying temporal coordination in black-bellied wren duets. *Animal Behaviour* 75: 1803–1808.

Lopez, J. K., and M. J. Fuxjager. 2012. Self-deception's adaptive value: effects of positive thinking and the winner effect. *Consciousness and Cognition* 21 (1): 315–324.

Loyau, A., D. Gomez, B. T. Moureau, M. Thery, N. S. Hart, M. Saint Jalme, A. T. D. Bennett, and G. Sorci. 2007. Iridescent structurally based coloration of eyespots correlates with mating success in the peacock. *Behavioral Ecology* 18 (6): 1123–1131.

Loyau, A., M. Petrie, M. Saint Jalme, and G. Sorci. 2008. Do peahens not prefer peacocks with more elaborate trains? *Animal Behaviour* 76: E5–E9.

Mac an Ghaill, M., and C. Haywood. 2012. Understanding boys: Thinking through boys, masculinity, and suicide. *Social Science and Medicine* 74 (4): 482–489.

Madsen, T., B. Ujvari, M. Olsson, and R. Shine. 2005. Paternal alleles enhance female reproductive success in tropical pythons. *Molecular Ecology* 14 (6): 1783–1787.

Madsen, T., and R. Shine. 1993. Male mating success and body size in European grass snakes. *Copeia* (2): 561–564.

Madsen, T., R. Shine, J. Loman, and T. Hakansson. 1993. Determinants of mating success in male adders, *Vipera berus*. *Animal Behaviour* 45 (3): 491–499.

Magnusson, W. E. 1979. Production of an embryo by an *Acrochordus javanicus* isolated for 7 years. *Copeia* (4): 744–745.

Maness, T. J., and D. J. Anderson. 2008. Mate rotation by female choice and coercive divorce in Nazca boobies, *Sula granti*. *Animal Behaviour* 76: 1267–1277.

Manning, G. J., C. J. Cole, H. C. Dessauer, and J. M. Walker. 2005. Hybridization between parthenogenetic lizards (*Aspidoscelis neomexicana*) and gonochoristic lizards (*Aspidoscelis sexlineata viridis*) in New Mexico: ecological, morphological, cytological, and molecular context. *American Museum Novitates* (3492): 3–56.

Marceau, C. 1985. When whiteness meant beauty. *Historia* (458): 88–96.

Marler, P., A. Dufty, and R. Pickert. 1986. Vocal communication in the domestic chicken: I. Does a sender communicate information about the quality of a food referent to a receiver? *Animal Behaviour* 34: 188–193.

Marlowe, F. W. 2004. Mate preferences among Hadza hunter-gatherers. *Human Nature* 15 (4): 365–376.

Marshall, A. J. 1954. Bower birds. *Biological Reviews* 29: 1–45.

Massen, J. J., A. M. Overduin-de Vries, A. J. de Vos-Rouweler, B. M. Spruijt, G. G. Doxiadis, and E. H. Sterck. 2012. Male mating tactics in captive rhesus macaques (*Macaca mulatta*): the influence of dominance, markets, and relationship quality. *International Journal of Primatology* 33 (1): 73–92.

Mattle, B., and A. B. Wilson. 2009. Body size preferences in the pot-bellied seahorse *Hippocampus abdominalis*: choosy males and indiscriminate females. *Behavioral Ecology and Sociobiology* 63 (10): 1403–1410.

Mazzi, D., R. Kunzler, and T. C. M. Bakker. 2003. Female preference for symmetry in computer-animated three-spined sticklebacks, *Gasterosteus aculeatus*. *Behavioral Ecology and Sociobiology* 54 (2): 156–161.

McCracken, K. G. 2000. The 20-cm spiny penis of the Argentine lake duck (*Oxyura vittata*). *Auk* 117 (3): 820–825.

Mehl, M. R., S. Vazire, N. Ramirez-Esparza, R. B. Slatcher, and J. W. Pennebaker. 2007. Are women really more talkative than men? *Science* 317 (5834): 82–82.

Mendez, V., R. D. Briceno, and W. G. Eberhard. 1998. Functional significance of the capitate supra-fronto-orbital bristles of male medflies (*Ceratitis capitata*) (Diptera, Tephritidae). *Journal of the Kansas Entomological Society* 71 (2): 164–174.

Meuche, I., O. Brusa, K. E. Linsenmair, A. Keller, and H. Prohl. 2013. Only distance matters: non-choosy females in a poison frog population. *Frontiers in Zoology* 10.

Milinski, M., D. Külling, and R. Kettler. 1990. Tit for tat: sticklebacks (*Gasterosteus aculeatus*) "trusting" a cooperating partner. *Behavioral Ecology* 1 (1): 7–11.

Milinski, M., I. Croy, T. Hummel, and T. Boehm. 2013. Major histocompatibility complex peptide ligands as olfactory cues in human body odour assessment. *Proceedings of the Royal Society B* 280 (1757).

Miller, E. J., and J. H. Kaufmann. 1975. Field observations of the social behaviour of the eastern grey kangaroo, *Macropus giganteu. Animal Behaviour* 23: 214–221.

Mitani, J. C. 1985. Location-specific responses of gibbons (*Hylobates muelleri*) to male songs. *Zeitschrift Fur Tierpsychologie* 70 (3): 219–224.

Moen, R., and J. Pastor. 1998. A model to predict nutritional requirements for antler growth in moose. *Alces* 34: 58–74.

Moller, A. P., M. Soler, and R. Thornhill. 1995. Breast asymmetry, sexual selection, and human reproductive success. *Ethology and Sociobiology* 16 (3): 207–219.

Moncho-Bogani, J., F. Martinez-Garcia, A. Novejarque, and E. Lanuza. 2005. Attraction to sexual pheromones and associated odorants in female mice involves activation of the reward system and basolateral amygdala. *European Journal of Neuroscience* 21 (8): 2186–2198.

Moore, A. J., W. J. Ciccone, M. D. Breed. 1988. The influence of social experience on the behavior of male cockroaches, *Nauphoeta cinerea. Journal of Insect Behavior* 1: 157–168.

Morris, P. H., J. White, E. R. Morrison, and K. Fisher. 2013. High heels as supernormal stimuli: how wearing high heels affects judgements of female attractiveness. *Evolution and Human Behavior* 34 (3): 176–181.

Mougeot, F., L. Perez-Rodriguez, N. Sumozas, and J. Terraube. 2009. Parasites, condition, immune responsiveness, and carotenoid-based ornamentation in male red-legged partridge *Alectoris rufa. Journal of Avian Biology* 40 (1): 67–74.

Muehlenhard, C. L., and S. K. Shippee. 2010. Men's and women's reports of pretending orgasm. *Journal of Sex Research* 47 (6): 552–567.

Mulder, R. A. 1997. Extra-group courtship displays and other reproductive tactics of superb fairy-wrens. *Australian Journal of Zoology* 45 (2): 131–143.

Muller, M. N., M. E. Thompson, and R. W. Wrangham. 2006. Male chimpanzees prefer mating with old females. *Current Biology* 16 (22): 2234–2238.

Navara, K. J., E. M. Anderson, and M. L. Edwards. 2012. Comb size and color relate to sperm quality: a test of the phenotype-linked fertility hypothesis. *Behavioral Ecology* 23 (5): 1036–1041.

Neave, N., and K. Shields. 2008. The effects of facial hair manipulation on female perceptions of attractiveness, masculinity, and dominance in male faces. *Personality and Individual Differences* 45 (5): 373–377.

Neufeld, C. J., and R. A. Palmer. 2012. Barnacle appendage plasticity: asymmetrical response time-lags, developmental mechanics, and seasonal variation. *Journal of Experimental Marine Biology and Ecology* 429: 20–27.

Oakwood, M. 2000. Reproduction and demography of the northern quoll, *Dasyurus hallucatus*, in the lowland savanna of northern Australia. *Australian Journal of Zoology* 48: 519–539.

Ostner, J., L. Vigilant, J. Bhagavatula, M. Franz, and O. Schülke. 2013. Stable heterosexual associations in a promiscuous primate. *Animal Behaviour* 86 (3): 632–631.

Owen-Smith, N. 1993. Comparative mortality rates of male and female kudus: the costs of sexual size dimorphism. *Journal of Animal Ecology* 62 (3): 428–440.

Oyegbile, T. O., and C. A. Marler. 2006. Weak winner effect in a less aggressive mammal: correlations with corticosterone but not testosterone. *Physiology and Behavior* 89 (2): 171–179.

Packer, C. 2010. Lions. *Current Biology* 20 (14): R590–R591.

Pagel, M., and W. Bodmer. 2003. A naked ape would have fewer parasites. *Proceedings of the Royal Society of London B* 270 (Suppl. 1): S117–S119.

Palagi, E. 2008. Sharing the motivation to play: the use of signals in adult bonobos. *Animal Behaviour* 75 (3): 887–896.

Parga, J. A. 2003. Copulatory plug displacement evidences sperm competition in *Lemur catta*. *International Journal of Primatology* 24 (4): 889–899.

Pariser, E. C., M. M. Mariette, and S. C. Griffith. 2010. Artificial ornaments manipulate intrinsic male quality in wild-caught zebra finches (*Taeniopygia guttata*). *Behavioral Ecology* 21 (2): 264–269.

Paul, E. L., and H. A. Hayes. 2002. The casualties of "casual sex": a qualitative exploration of the phenomenology of college students' hookups. *Journal of Social and Personal Relationships* 19 (5): 639–661.

Peixoto, P. E. C., and W. W. Benson. 2012. Influence of previous residency and body mass in the territorial contests of the butterfly *Hermeuptychia fallax* (Lepidoptera: Satyrinae). *Journal of Ethology* 30 (1): 61–68.

Penn, D. J., and W. K. Potts. 1999. The evolution of mating preferences and major histocompatibility complex genes. *American Naturalist* 153 (2): 145–164.

Perlesz, A., J. Power, R. Brown, R. McNair, M. Schofield, M. Pitts, A. Barrett, and A. Bickerdike. 2010. Organising work and home in same-sex parented families: findings from the Work Love Play Study. *Australian and New Zealand Journal of Family Therapy* 31 (4): 374–391.

Peshek, D., N. Semmaknejad, D. Hoffman, and P. Foley. 2011. Preliminary evidence that the limbal ring influences facial attractiveness. *Evolutionary Psychology* 9 (2): 137–146.

Peterkin, A. 2002. One thousand beards: a cultural history of facial hair. Vancouver, BC: Arsenal Pulp Press.

Petersson, E., and T. Jarvi. 2001. "False orgasm" in female brown trout: trick or treat? *Animal Behaviour* 61: 497–501.

Petrie, M., P. Cotgreave, and T. W. Pike. 2009. Variation in the peacock's train shows a genetic component. *Genetica* 135 (1): 7–11.

Pfefferle, D., P. M. West, J. Grinnell, C. Packer, and J. Fischer. 2007. Do acoustic features of lion, *Panthera leo*, roars reflect sex and male condition? *Journal of the Acoustical Society of America* 121 (6): 3947–3953.

Pflüger, L. S., E. Oberzaucher, S. Katina, I. J. Holzleitner, and K. Grammer. 2012. Cues to fertility: perceived attractiveness and facial shape predict reproductive success. *Evolution and Human Behavior* 33 (6): 708–714.

Pickering, S. P. C., and S. D. Berrow. 2001. Courtship behaviour of the wandering albatross *Diomedea exulans* at Bird Island, South Georgia. *Marine Ornithology* 29: 29–37.

Pines, M., and L. Swedell. 2011. Not without a fair fight: failed abductions of females in wild hamadryas baboons. *Primates* 52 (3): 249–252.

Pisanski, K., and D. R. Feinberg. 2013. Cross-cultural variation in mate preferences for averageness, symmetry, body size, and masculinity. *Cross-Cultural Research* 47 (2): 162–197.

Pitcher, T. E., B. D. Neff, F. H. Rodd, and L. Rowe. 2003. Multiple mating and sequential mate choice in guppies: females trade up. *Proceedings of the Royal Society B* 270 (1524): 1623–1629.

Place, S. S., P. M. Todd, L. Penke, and J. B. Asendorpf. 2010. Humans show mate copying after observing real mate choices. *Evolution and Human Behavior* 31 (5): 320–325.

Povoledo, E., and D. Carvajal. 2012. Increasingly in Europe, suicides "by economic crisis." *New York Times*, April 14.

Price, Cathy J. 2012. A review and synthesis of the first 20 years of PET and fMRI studies of heard speech, spoken language, and reading. *NeuroImage* 62 (2): 816–847.

Procter, D. S., A. J. Moore, and C. W. Miller. 2012. The form of sexual selection arising from male-male competition depends on the presence of females in the social environment. *Journal of Evolutionary Biology* 25 (5): 803–812.

Prokop, Z. M., L. Michalczyk, S. M. Drobniak, M. Herdegen, and J. Radwan. 2012. Meta-analysis suggests choosy females get sexy sons more than "good genes." *Evolution* 66 (9): 2665–2673.

Prokosch, M. D., R. G. Coss, J. E. Scheib, and S. A. Blozis. 2009. Intelligence and mate choice: intelligent men are always appealing. *Evolution and Human Behavior* 30: 11–20.

Provost, M. P., N. F. Troje, and V. L. Quinsey. 2008. Short-term mating strategies and attraction to masculinity in point-light walkers. *Evolution and Human Behavior* 29 (1): 65–69.

Pruitt, J. N., and S. E. Riechert. 2012. The ecological consequences of temperament in spiders. *Current Zoology* 58 (4): 589–596.

Pryke, S. R., and S. Andersson. 2005. Experimental evidence for female choice and energetic costs of male tail elongation in red-collared widowbirds. *Biological Journal of the Linnean Society* 86 (1): 35–43.

Purcell, R., M. Pathe, and P. E. Mullen. 2001. A study of woman who stalk. *American Journal of Psychiatry* 158 (12): 2056–2060.

Puts, D. A. 2010. Beauty and the beast: mechanisms of sexual selection in humans. *Evolution and Human Behavior* 31 (3): 157–175.

Qvarnström, A., V. Blomgren, C. Wiley, and N. Svedin. 2004. Female collared flycatchers learn to prefer males with an artificial novel ornament. *Behavioral Ecology* 15 (4): 543–548.

Ramachandran, V. S., and S. Blakeslee. 1999. *Phantoms in the Brain: Probing the Mysteries of the Human Mind.* New York: William Morrow, pp. 35–37.

Ratliff, K. A., and S. Oishi. 2013. Gender differences in implicit self-esteem following a romantic partner's success or failure. *Journal of Personality and Social Psychology* 105 (4): 688–702.

Réale, D., P. Boussas, and J. L. Chapuis. 1996. Female-biased mortality induced by male sexual harassment in a feral sheep population. *Canadian Journal of Zoology* 74 (10): 1812–1818.

Reaney, L. T., J. M. Drayton, and M. D. Jennions. 2011. The role of body size and fighting experience in predicting contest behaviour in the black field cricket, *Teleogryllus commodus*. *Behavioral Ecology and Sociobiology* 65 (2): 217–225.

Reichert, M. S. 2010. Aggressive thresholds in *Dendropsophus ebraccatus*: habituation and sensitization to different call types. *Behavioral Ecology and Sociobiology* 64 (4): 529–539.

Reichert, M. S., and H. C. Gerhardt. 2011. The role of body size on the outcome, escalation, and duration of contests in the grey treefrog, *Hyla versicolor*. *Animal Behaviour* 82 (6): 1357–1366.

Ribble, D. O. 1991. The monogamous mating system of *Peromyscus californicus* as revealed by DNA fingerprinting. *Behavioral Ecology and Sociobiology* 29 (3): 161–166.

Robbins, M. B., and Jensen, G. D. 1978. Multiple orgasm in males. *Journal of Sex Research* 14 (1): 21–26.

Robel, R. J. 1966. Booming territory size and mating success of the greater prairie chicken (*Tympanuchus cupido pinnatus*). *Animal Behaviour* 14 (2): 328–331.

Roberts, S. C., A. Kralevich, C. Ferdenzi, T. K. Saxton, B. C. Jones, L. M. DeBruine, A. C. Little, and J. Havlicek. 2011. Body odor quality predicts behavioral attractiveness in humans. *Archives of Sexual Behavior* 40 (6): 1111–1117.

Roberts, S. C., L. M. Gosling, C. Vaughan, and M. Petrie. 2008. MHC-correlated odour preferences in humans and the use of oral contraceptives. *Proceedings of the Royal Society B* 275: 2715–2722.

Rodríguez-Muñoz, R., A. Bretman, and T. Tregenza. 2011. Guarding males protect females from predation in a wild insect. *Current Biology* 21 (20): 1716–1719.

Roldan, E. R. S., J. Cassinello, T. Abaigar, and M. Gomendio. 1998. Inbreeding, fluctuating asymmetry, and ejaculate quality in an endangered ungulate. *Proceedings of the Royal Society B* 265: 243–248.

Roney, J. R., and Z. L. Simmons. 2012. Men smelling women: null effects of exposure to ovulatory sweat on men's testosterone. *Evolutionary Psychology* 10 (4): 703–713.

Rotjan, R. D., J. R. Chabot, and S. M. Lewis. 2010. Social context of shell acquisition in *Coenobita clypeatus* hermit crabs. *Behavioral Ecology* 21 (3): 639–646.

Sacco, D. F., S. G. Young, C. M. Brown, M. J. Bernstein, and K. Hugenberg. 2012. Social exclusion and female mating behavior: rejected women show strategic enhancement of short-term mating interest. *Evolutionary Psychology* 10 (3): 573–587.

Scharf, I., and O. Y. Martin. 2013. Same-sex sexual behavior in insects and arachnids: prevalence, causes, and consequences. *Behavioral Ecology and Sociobiology* 67 (11): 1719–1730.

Scheib, J. E., S. W. Gangestad, and R. Thornhill. 1999. Facial attractiveness, symmetry and, cues of good genes. *Proceedings of the Royal Society B* 266 (1431): 1913–1917.

Schiestl, F. P., and S. Cozzolino. 2008. Evolution of sexual mimicry in the orchid subtribe orchidinae: the role of preadaptations in the attraction of male bees as pollinators. *BMC Evolutionary Biology* 8.

Schiller, D., J. B. Freeman, J. P. Mitchell, J. S. Uleman, and E. A. Phelps. 2009. A neural mechanism of first impressions. *Nature Neuroscience* 12 (4): 508–514.

Schmitt, D. P., T. K. Shackleford, J. Duntley, W. Tooke, and D. M. Buss. 2001. The desire for sexual variety as a key to understanding basic human mating strategies. *Personal Relationships* 8 (4): 425–455.

Schuett, G. W. 1997. Body size and agonistic experience affect dominance and mating success in male copperheads. *Animal Behaviour* 54: 213–224.

Schwagmeyer, P. L., and S. J. Woontner. 1986. Scramble competition polygyny in thirteen-lined ground squirrels: the relative contributions of overt conflict and competitive mate searching. *Behavioral Ecology and Sociobiology* 19 (5): 359–364.

Scott, D. K. 1980. Functional aspects of the pair bond in winter in Bewick's swans (*Cygnus columbianus bewickii*). *Behavioral Ecology and Sociobiology* 7 (4): 323–327.

Sefcek, J. A., and J. E. King. 2007. Chimpanzee facial symmetry: a biometric measure of chimpanzee health. *American Journal of Primatology* 69 (11): 1257–1263.

Semlitsch, R. D. 1994. Evolutionary consequences of nonrandom mating: do large males increase offspring fitness in the anuran *Bufo bufo*. *Behavioral Ecology and Sociobiology* 34 (1): 19–24.

Semple, S., and K. McComb. 1996. Behavioural deception. *Trends in Ecology and Evolution* 11 (10): 434–437.

Serrano-Meneses, M. A., A. Cordoba-Aguilar, V. Mendez, S. J. Layen, and T. Szekely. 2007. Sexual size dimorphism in the American rubyspot: male body size predicts male competition and mating success. *Animal Behaviour* 73: 987–997.

Setchell, J. M., and E. J. Wickings. 2006. Mate choice in male mandrills (*Mandrillus sphinx*). *Ethology* 112 (1): 91–99.

Shackelford, T. K., and A. T. Goetz. 2007. Adaptation to sperm competition in humans. *Current Directions in Psychological Science* 16 (1): 47–50.

Shealer, D. A., J. A. Spendelow, J. S. Hatfield, and I. C. T. Nisbet. 2005. The adaptive significance of stealing in a marine bird and its relationship to parental quality. *Behavioral Ecology* 16 (2): 371–376.

Shelton, B. A., and D. John. 1996. The division of household labor. *Annual Review of Sociology* 22: 299–322.

Sherman, P. W. 1998. Animal behavior: the evolution of menopause. *Nature* 392 (6678): 759–61.

Shine, R., C. Gordon, and G. Terri. 1981. Mating and male combat in Australian blacksnakes, *Pseudechis porphyriacus*. *Journal of Herpetology* 15 (1): 101–107.

Shuster, S. M., and M. J. Wade. 1991. Equal mating success among male reproductive strategies in a marine isopod. *Nature* 350 (6319): 608–610.

Silk, J. B., P. S. Rodman, and A. Samuels. 1979. Kidnapping and spite in female infant relations of bonnet macaques. *American Journal of Physical Anthropology* 50 (3): 481–482.

Silk, J. B., R. M. Seyfarth, and D. L. Cheney. 1999. The structure of social relationships among female savanna baboons in Moremi Reserve, Botswana. *Behaviour* 136: 679–703.

Simon, Robin W., and Anne E. Barrett. 2010. Nonmarital romantic relationships and mental health in early adulthood: does the association differ for women and men? *Journal of Health and Social Behavior* 51 (2): 168–182.

Singh, D., B. J. Dixson, T. S. Jessop, B. Morgan, and A. F. Dixson. 2010. Cross-cultural consensus for waist-hip ratio and women's attractiveness. *Evolution and Human Behavior* 31 (3): 176–181.

Smith, C. L., D. A. Van Dyk, P. W. Taylor, and C. S. Evans. 2009. On the function of an enigmatic ornament: wattles increase the conspicuousness of visual displays in male fowl. *Animal Behaviour* 78 (6): 1433–1440.

Smuts, B. B. 1985. *Sex and Friendship in Baboons*. Hawthorne, NY: Aldine.

Smuts, B. B., and R. W. Smuts. 1993. Male aggression and sexual coercion of females in nonhuman primates and other mammals: evidence and theoretical implications. *Advances in the Study of Behavior* 22: 1–63.

Sokhi, D. S., et al. 2005. Men hear women's melodies: male and female voices activate distinct regions in the male brain. *NeuroImage* 27 (3): 572–578.

Speck, F. G., and J. Witthoft. 1947. Some notable life-histories in zoological folklore. *Journal of American Folklore* 60 (238): 345–349.

Spoon, T. R., J. R. Millam, and D. H. Owings. 2006. The importance of mate behavioural compatibility in parenting and reproductive success by cockatiels, *Nymphicus hollandicus*. *Animal Behaviour* 71 (2): 315–326.

Squires, K. A., K. Martin, and R. I. Goudie. 2007. Vigilance behavior in the harlequin duck (*Histrionicus histrionicus*) during the preincubation period in Labrador: are males vigilant for self or social partner? *Auk* 124 (1): 242–252.

Stanik, C. E., and P. C. Ellsworth. 2010. Who cares about marrying a rich man? Intelligence and variation in women's mate preferences. *Human Nature* 21: 203–217.

Starkweather, K. E., and R. Hames. 2012. A survey of non-classical polyandry. *Human Nature* 23 (2): 149–172.

Stewart, S., H. Stinnett, and L. B. Rosenfeld. 2000. Sex differences in desired characteristics of short-term and long-term relationship partners. *Journal of Social and Personal Relationships* 17 (6): 843–853.

Straus, M. A. 2004. Prevalence of violence against dating partners by male and female university students worldwide. *Violence against Women* 10 (7): 790–811.

Sundaresan, S. R., I. R. Fischhoff, and D. I. Rubenstein. 2007. Male harassment influences female movements and associations in Grevy's zebra (*Equus grevyi*). *Behavioral Ecology* 18 (5): 860–865.

Sunnafrank, M., and A. Ramirez. 2004. At first sight: persistent relational effects of get-acquainted conversations. *Journal of Social and Personal Relationships* 21 (3): 361–379.

Svensson, P. A., T. K. Lehtonen, and B. B. M. Wong. 2012. A high aggression strategy for smaller males. *PLOS ONE* 7 (8): e43121.

Swami, V., R. Miller, A. Furnham, L. Penke, and M. J. Tovée. 2008. The influence of men's sexual strategies on perceptions of women's bodily attractiveness, health, and fertility. *Personality and Individual Differences* 44 (1): 98–107.

Tecot, S. R., B. D. Gerber, S. J. King, J. L. Verdolin, and P. C. Wright. 2013. Risky business: sex differences in mortality and dispersal in a polygynous, monomorphic lemur. *Behavioral Ecology* 24 (4): 987–996.

Tedore, C., and S. Johnsen. 2012. Weaponry, color, and contest success in the jumping spider *Lyssomanes viridis*. *Behavioural Processes* 89 (3): 203–211.

Thacker, P. D. 2004. Biological clock ticks for men, too: genetic defects linked to sperm of older fathers. *Journal of the American Medical Association* 291 (14): 1683–1685.

Thayer, Z. M., and S. D. Dobson. 2013. Geographic variation in chin shape challenges the universal facial attractiveness hypothesis. *PLOS ONE* 8 (4).

Thomas, D. W., M. B. Fenton, and R. M. Barclay. 1979. Social behavior of the little brown bat, *Myotis lucifugus*. *Behavioral Ecology and Sociobiology* 6 (2): 129–136.

Thomas, M. L. 2011. Detection of female mating status using chemical signals and cues. *Biological Reviews* 86 (1): 1–14.

Thornhill, R., and K. Grammer. 1999. The body and face of woman: one ornament that signals quality? *Evolution and Human Behavior* 20 (2): 105–120.

Thornhill, R., and S. W. Gangestad. 1999. The scent of symmetry: a human sex pheromone that signals fitness? *Evolution and Human Behavior* 20 (3): 175–201.

———. 2006. Facial sexual dimorphism, developmental stability, and susceptibility to disease in men and women. *Evolution and Human Behavior* 27 (2): 131–144.

Troisi, A., and M. Carosi. 1998. Female orgasm rate increases with male dominance in Japanese macaques. *Animal Behaviour* 56: 1261–1266.

Uchida, N., A. Kepecs, and Z. F. Mainen. 2006. Seeing at a glance, smelling in a whiff: rapid forms of perceptual decision making. *Nature Reviews Neuroscience* 7 (6): 485–491.

Valero, A., C. M. Garcia, and A. E. Magurran. 2008. Heterospecific harassment of native endangered fishes by invasive guppies in Mexico. *Biology Letters* 4 (2): 149–152.

Van der Jeugd, H. P., and K. B. Blaakmeer. 2001. Teenage love: the importance of trial liaisons, subadult plumage, and early pairing in barnacle geese. *Animal Behaviour* 62 (6): 1075–1083.

Van Dongen, S. 2012. Fluctuating asymmetry and masculinity/femininity in humans: a meta-analysis. *Archives of Sexual Behavior* 41 (6): 1453–1460.

VanderWaal, K. L., A. Mosser, and C. Packer. 2009. Optimal group size, dispersal decisions, and postdispersal relationships in female African lions. *Animal Behaviour* 77 (4): 949–954.

Van Dongen, S., R. Cornille, and L. Lens. 2009. Sex and asymmetry in humans: what is the role of developmental instability? *Journal of Evolutionary Biology* 22 (3): 612–622.

Vehrencamp, S. L. 1983. A model for the evolution of despotic versus egalitarian societies. *Animal Behaviour* 31 (August): 667–682.

Verburgt, L., M. Ferreira, and J. W. H. Ferguson. 2011. Male field cricket song reflects age, allowing females to prefer young males. *Animal Behaviour* 81 (1): 19–29.

Vingilis-Jaremko, L., and D. Maurer. 2013. The influence of symmetry on children's judgments of facial attractiveness. *Perception* 42 (3): 302–320.

Voracek, M., M. Fisher, and T. K. Shackelford. 2009. Sex differences in subjective estimates of non-paternity rates in Austria. *Archives of Sexual Behavior* 38 (5): 652–656.

Wade, T. J. 2010. The relationships between symmetry and attractiveness and mating relevant decisions and behavior: a review. *Symmetry* 2: 1081–1098.

Waits, D. P. 1989. Infanticide in mountain gorillas: new cases and a reconsideration of the evidence. *Ethology* 81:1–18.

Waitt, C., and A. C. Little. 2006. Preferences for symmetry in conspecific facial shape among *Macaca mulatta*. *International Journal of Primatology* 27 (1): 133–145.

Walker, M. L., and J. G. Herndon. 2008. Menopause in nonhuman primates? *Biology of Reproduction* 79 (3): 398–406.

Walum, H., L. Westberg, S. Henningsson, J. M. Neiderhiser, D. Reiss, W. Igl, J. M. Ganiban, E. L. Spotts, N. L. Pedersen, E. Eriksson, and P. Lichtenstein. 2008. Genetic variation in the vasopressin receptor 1a gene (AVPR1A) associates with pair-bonding behavior in humans. *Proceedings of the National Academy of Sciences* 105 (37): 14153–14156.

Waynforth, D. 2007. Mate choice copying in humans. *Human Nature* 18 (3): 264–271.

Webberley, K. M., J. Buszko, V. Isham, and G. D. D. Hurst. 2006. Sexually transmitted disease epidemics in a natural insect population. *Journal of Animal Ecology* 75 (1): 33–43.

Wedekind, C., and D. Penn. 2000. MHC genes, body odours, and odour preferences. *Nephrology Dialysis Transplantation* 15 (9): 1269–1271.

Wedell, N., and M. G. Ritchie. 2004. Male age, mating status, and nuptial gift quality in a bushcricket. *Animal Behaviour* 67: 1059–1065.

Wedell, N., M. J. Gage, and G. A. Parker. 2002. Sperm competition, male prudence, and sperm-limited females. *Trends in Ecology and Evolution* 17 (7): 313–320.

Welling, L. L. M., L. M. DeBruine, A. C. Little, and B. C. Jones. 2009. Extraversion predicts individual differences in women's face preferences. *Personality and Individual Differences* 47 (8): 996–998.

West, P. M., and C. Packer. 2002. Sexual selection, temperature, and the lion's mane. *Science* 297 (5585): 1339–1343.

Whithey, A. 2013. Shaving and masculinity in eighteenth-century Britain. *Journal of Eighteenth-Century Studies* 36 (2): 225–243.

Willis, J., and A. Todorov. 2006. First impressions: making up your mind after a 100-ms exposure to a face. *Psychological Science* 17 (7): 592–598.

Winslow, J. T., N. Hastings, C. S. Carter, C. R. Harbaugh, and T. R. Insel. 1993. A role for central vasopressin in pair bonding in monogamous prairie voles. *Nature* 365 (6446): 545–548.

Witte, K., and B. Caspers. 2006. Sexual imprinting on a novel blue ornament in zebra finches. *Behaviour* 143: 969–991.

Wolff, J. O., and D. W. Macdonald. 2004. Promiscuous females protect their offspring. *Trends in Ecology and Evolution* 19 (3): 127–134.

Woodroffe, R., and D. W. Macdonald. 1995. Female/female competition in European badgers *Meles meles*: effects on breeding success. *Journal of Animal Ecology* 64 (1): 12–20.

Yamada, T., K. Hara, H. Umematsu, and T. Kadowaki. 2013. Male pattern baldness and its association with coronary heart disease: a meta-analysis. *BMJ Open* 3.

Yasukawa, K., and W. A. Searcy. 1982. Aggression in female red-winged blackbirds: a strategy to ensure male parental investment. *Behavioral Ecology and Sociobiology* 11 (1): 13–17.

Yoda, K., and Y. Ropert-Coudert. 2002. A short note on an Adélie penguin feeding its own mate. *Polar Biology* 25 (11): 868–869.

Zuk, M., R. Thornhill, J. D. Ligon, K. Johnson, S. Austad, S. H. Ligon, N. W. Thornhill, and C. Costin. 1990. The role of male ornaments and courtship behavior in female mate choice of red jungle fowl. *American Naturalist* 136 (4): 459–473.

INDEX